TOPICS IN THE FOUNDATION OF STATISTICS

Edited by

Bas C. van Fraassen

Department of Philosophy,
Princeton University

Reprinted from
Foundations of Science
Volume 1, No. 1, 1995/96

Springer-Science+Business Media, B.V.

A C.I.P. Catalogue record for this book is available from the Library of Congress

ISBN 978-90-481-4792-2   ISBN 978-94-015-8816-4 (eBook)
DOI 10.1007/978-94-015-8816-4

Printed on acid-free paper

Prepared with permission of Oficyna Akademicka

CONTENTS

Foundations of Science
1 (1995/96), 5-18

FOUNDATIONS OF SCIENCE – DISCUSSION

Bas van Fraassen
Department of Philosophy
Princeton University
Princeton, NJ 08544, USA

A PHILOSOPHICAL APPROACH TO FOUNDATIONS OF SCIENCE

Abstract. Foundational research focuses on the theory, but theories are to be related also to other theories, experiments, facts in their domains, data, and to their uses in applications, whether of prediction, control, or explanation. A theory is to be identified through its class of models, but not so narrowly as to disallow these roles. The language of science is to be studied separately, with special reference to the relations listed above, and to the consequent need for resources other than for theoretical description. Peculiar to the foundational level are questions of completeness (specifically in the representation of measurement), and of interpretation (a topic beset with confusions of truth and evidence, and with inappropriate metalinguistic abstraction).

In foundational research, focus is on the theory, that is, on the product of theoretical activity in science, as opposed to, for example, on the processes of theory testing, choice, evaluation, confirmation, or historic and other contextual factors that may play a role in such theorizing.

Nevertheless, such research pertains not only to the structure of the theory itself, but also to its relations to other theories, to the facts in the domain of the theory, to experiment, to data obtained from experiment and observation, and to its use in applications, whether of prediction, control, or explanation.

1. **What is a theory?** Here I am in agreement with Patrick Suppes, that we must distinguish the theory from its formulations. Thus a set of axioms or theorems cannot be the theory; the theory is something which admits

of many alternative formulations, which may indeed be given in alternative languages with different vocabularies and even different logical resources. For this reason it is is best to identify a theory through a class of mathematical structures, its models, which can be described in those various ways.

At the same time, however, a theory gives information about the world, and may be believed or doubted. Therefore, although a theory may be identified through a class of models, and may even (in accordance with logical and mathematical practice) be identified with that class, there is clearly more to it. For a class is not the sort of thing that gives information or may be believed or doubted.

Historically, this point has sometimes been made by saying that scientific theories are not "uninterpreted objects" such as are studied in pure mathematics. That is not a good choice of words, it seems to me, and may be misleading. It suggests that a scientific theory could be instead a composite, something abstract together with something else that interprets that abstract thing. But then what is that something else? Is it too something abstract, like a function? If so we have the same problem all over again – we have replaced one abstract thing by two. On the other hand if it is suggested that a function needs no interpretation, though it is an abstract thing, then why did the first abstract thing need interpretation? If on the third hand the something else is not an abstract thing at all, but something concrete, we shall have to study it (scientifically, in the way all concrete things are to be studied) – and how shall we do that, if not by constructing a mathematical model of it?

Therefore I prefer to say that a theory can be identified *through* its class of models. These models are the main subject of study for foundational research, taken in its strictest sense now, pertaining to that theory. But the theory states that these models include a correct representation of what there is (or of what the theory is about). This statement (by which I do not mean any linguistic formulation, but rather what all formulations of the theory "say") is subject to assent or doubt. It is also subject to interpretation prior to any assent or doubt. To this role of interpretation I shall return below.

We must also note here the importance and propriety of Suppes' call to make mathematics rather than meta-mathematics the tool of philosophy of science with respect to foundational discussions. The class of models is a class of mathematical structures, described in any way mathematics allows. Mathematical description has its limits of course, in that mathematics describes its subject matter only up to isomorphism.

2. Logical positivism and logical empiricism began, somewhat unfortu-

nately, by assimilating the relation between a theory and its domain to the relation of theory to evidence. Both are important, and to some extent they overlap. Those relations of theory to experiment that fall under such epistemic headings as confirmation and testing I do not count as belonging to the area of foundations.

However, when we consider the relation of the theory to fact (to its domain, to the world, however we want to phrase this), it is important to ask whether the theory itself models the processes of measurement or experiment, or whether those are described in terms of some other theory. This becomes crucial for those theories in physics that are potentially universal theories, theories of everything, such as quantum mechanics and general relativity. When the adequacy of the theory requires it to have models which correctly represent certain phenomena, and this adequacy can be ascertained only through certain types of processes, we face a potential consistency problem. Does the theory represent not only those phenomena but also those processes which give experimental access to those phenomena? If so, do the two sorts of representations 'mesh'? This is the form of the measurement problem in quantum mechanics, but may crop up in any foundational discussion, in some form or other.

3. Uses of a theory comprise prediction and explanation. Both of these are topics that tend to lead us from foundations into problems concerning the language of science. Prediction requires the user to relate himself to some model or models of the theory, in a way that is analogous to someone who specifies his own location on a map, or gives the co-ordinates of his location, or specifies the co-ordinate system which is 'rigidly attached' to him. The linguistic resources to do this do not belong to the purely theoretical language which is sufficient for the presentation of the theory, but to the language of applied science. Since the description of the models is mathematical, this co-ordination with the theory required for an application may involve more than specifying the user's own co-ordinates. For example, he may wish to use a certain formal theory to predict thermal phenomena, while the equations of that theory describe equally well e. g. gas diffusion.

It seems to me that Toulmin's analogy between a theory and a map is quite apt here, even if he used it differently. A circuit diagram might, by coincidence, share some structure with a roadmap or railway map, or a map of one city might share some structure with that of another. In that case I can use one in place of the other, by ignoring its origin, paying attention only to the relevant structure, and locating myself on it in a different way. There is a certain limitation of language that is evident here, where theory and practice meet. For describing the act of self-location on the map is very different from doing it, and until it is done, conditions for the use of the map are not met.

The use of a theory in explanation may require even more, at least if accounts of explanation in terms of the logic of questions are correct. Issues of completeness for theories – such as the much discussed completeness of quantum mechanics issue – relate solely to the expressive power of the theoretical language, i. e. the language in which the models and the domain of the theory are described. Use of the theoretical description as explanation, however, may require attention to very specific interests and concerns characterizing the context of inquiry. The very same description may be able to play the role of explanation in one context and not in another – so explanatory completeness does not follow from expressive completeness.

4. Some of the problems touched on under the above headings do not pertain to foundations of the sciences at all, although they belong to philosophy of science or in some cases to science. In addition, some of the problems indicated were left dangling, so to speak. I will here add comments only about one. Among the relations of a theory to its domain there are those which determine whether or not the theory is true. Prior to the question of truth, however, are questions of interpretation and of possibility. The question of truth does not arise except relative to an interpretation. On a purely theoretical level we can investigate which interpretations the theory admits, and thereby answer the question: how could the world (domain, fact) possibly be the way this theory says it is? That it is not easy to answer this question in general is clearly due to two factors. The first is the difficulty of developing any interpretation at all for sufficiently complex theories presented in mathematical form. The second is the further demand that the question be answered through an interpretation on which the theory is compatible with other accepted parts of science. In the case of quantum mechanics, for example, we have found that it admits no interpretations which cohere with certain traditional assumptions about causality, while on the other hand, interpretations that imply the existence of superluminal signals are not generally considered satisfactory either.

In conclusion I would like to draw attention to ongoing work to elaborate on the conception of foundational research in the sense of section (1) above, which received its original impetus from the work of Suppes. Some history is provided by way of introduction in Fred Suppe (Suppe, 1989); a recent contribution to be noted is (da Costa and French, 1990).

References

Da Costa, N.C.A. and French, S. (1990), The model-theoretic approach in the philosophy of science, *Philosophy of Science* **57**, 248-265.

Suppe, F. (1989), *The Semantic Conception of Theories and Scientific Realism*. Urbana: University of Illinois Press.

Patrick Suppes
Stanford University
Stanford, CA 943005, USA

A PLURALISTIC VIEW OF FOUNDATIONS OF SCIENCE

Before writing this I read Bas van Fraassen's statement. I find little I can disagree with in what he says. Consequently, what I would like to do is address several clusters of concepts and issues that are often somewhat neglected in discussions of the philosophical foundations of science.

I. Epistemology of Experiments

Bas rightly emphasizes the importance of going beyond models of theories in analyzing the content of science. What I want to stress is how far the whole activity of experimentation, including its reporting, is from the standard theoretical model of science, which Bas and I substantially agree about.

The first point is that experimental reports, much more than theoretical papers, are not generally understandable. They are like detailed sports reporting, intelligible only to the initiated. There is, of course, the insistence of journals that reports be as brief as possible, and certainly there is no requirement they be generally intelligible. However, I think there is a deeper epistemological reason, which also applies to sports reporting. It is not possible to describe in ordinary language, even augmented by some technical terms, the backhand stroke of a tennis player with any accuracy. Similarly, it is not possible to describe in language the many activities of an experimentalist. This applies to his actions setting up and running experiments, but also to his perceptions in digesting the results in their first "raw" form.The written account can only hint at the main features of what is done or what is observed.

The central epistemological point of these remarks is that this radical incompleteness of descriptions of experiments is not a mark of bad science, but is an essential, unremovable feature of almost all science. The drastic descriptive limitation of what we have to say about experiments is, in my view, a fundamental limitation of our scientific knowledge, possible or actual.

Moreover, this radical incompleteness of the experimental reporting leads to more appeals to authority in experimental work than in theoretical work. It is common to hear, in every part of experimental science: "Well, we know those results are right and can be trusted because we know X and his lab." Of course, there is the answer that experiments can be repeated by others, and this is the great empirical check against being overwhelmed by authority. But it is still an important point, which can be easily amplified by various historical examples, that in many respects theoretical results can be evaluated for error much easier and more directly than experimental work. To bring this up to current science, computational experiments and simulations need to be included as well.

Part of the epistemology of experiments recognized by everyone is the presence of experimental error, but the theory of error has not crept into the philosophical foundations of science, but remains on the ground floor of actual experimental work, with only an occasional philosophical nod to its importance. Yet error is a central concept of a proper general epistemology, and, on the other hand, has a long technical history of theoretical development, at least since the early work of Simpson in the 18th century. Its conceptual place in science, however, is, in my view, still not fully recognized. Let me give just one personal example. I have spent many years working on the foundations of measurement, and I recently went to a gathering of the measurement theorists' "clan" in Kiel, Germany. It was generally agreed that a really proper inclusion and analysis of error in foundational theories of measurement is the number one general problem.

Finally, I emphasize, as I have before, the hierarchy of models used in the analysis of experimental data. Modern statistics has developed within the set-theoretical view of mathematics, as can easily be seen by perusing the pages of *The Annals of Statistics* and other leading journals. The many levels of data reduction usually needed to get to detailed statistical analysis is an epistemological problem, not a statistical problem as such, and needs more philosophical analysis, with close attention to the varying practices in different parts of science. Skepticism about explicit objectivity holding "all the way down" is certainly one of the reasons for the recent increase in acceptance of a Bayesian viewpoint in statistics, although it was already very explicit at many points in Laplace's writing. The arguments about Bayesian ideas, both by philosophers and statisticians, are highly relevant to the epistemology of experiments.

II. Theoretical Physics is not Primitive Mathematics

I continue to endorse the set-theoretical view of theories as an important foundational viewpoint, but I think there is much about theoretical science that is left out of this view. I especially have in mind the detailed comparison of the methods of theoretical physics and the methods of pure mathematics. The latter come close to satisfying the set-theoretical view of theories. The former do not. It is possible to take the view that it is the job of philosophers of physics to fill in the set-theoretical foundations of physics. (Certainly it is the kind of enterprise I have engaged in myself in the past, and it is not my point here to denigrate it.) What I want to emphasize here is that it is also philosophically important to understand the difference in the way mathematical methods are used in theoretical physics and in pure mathematics, or for that matter, standard applied mathematics – from the standpoint of the discussion here I make no real distinction between the practices of pure and applied mathematics.

Even the most superficial empirical examination of *Physics Review Letters*, the journal that perhaps more than any other publishes important new physics results, both experimental and theoretical, will confirm what I have to say, or at least so I claim. The articles, strictly limited as to length, do not read at all like articles in mathematics journals. Mathematics is used in an informal way, but that phrase is too vague. The real characterization, I think, is that theoretical physics is concerned with problem solving, in strong contrast to theorem proving. The essential skill is knowing just what assumptions or empirical facts to draw upon. Formal axioms seldom if ever play a role.

I should say at once that mathematical physics, as opposed to theoretical physics, does use axioms. Look at the heroic efforts to provide an axiomatic foundation for quantum field theory. Essentially all the sustained efforts to provide such an analysis were made by mathematicians, not theoretical physicists.

Perhaps my favorite exhibits for my thesis about the difference are the papers of the two great magicians of 20th century physics, Dirac and Feynman. Any fundamental paper by either one of them is far from being mathematically complete, either in terms of assumptions or argument. But these are only wonderful examples. I think the difference holds for almost any of the first theoretical papers in any part of modern physics. (I say "almost any" although I believe "all", because I have not looked at a wide enough range of candidates.)

The saliency of the methodological differences between theoretical physics and mathematics is even more evident in physical papers and treatises further removed from what we all tend to regard as fundamental theory. For example, papers and books on quantum optics or solid state physics are full of mathematical equations and mathematical calculations. I say "calculations" deliberately, for it roughly characterizes the restricted kind of mathematical arguments used. One can look from one physical paper or book to the next and no trace of even an elementary reductio ad absurdum argument or of a pure exisence proof will be found. (Undoubtedly isolated counter examples can be found to what I am saying, but they will be very isolated, i.e., of very low frequency.)

So the three main characteristics of the use of mathematics in theoretical physics are these. First, they are highly constructive, mainly calculational in character – the calculus is not called the calculus by some inadvertent mistake. Second, they are fragmentary, a calculation, then a physical principle or argument, then another calculation, all weaved together into something physically plausible but far from being an exemplar of purely mathematical argument. Third, the broad organization of the argument fits a problem-solving not theorem-proving style. It is not that theoretical physicists forget to prove the theorems, or don't know how to, it is that they are doing something different, something that is in its own way just as wonderful and impressive as sustained proofs in mathematics.

It is a problem for the philosophy of science to understand the detailed nature of the methods of theoretical physics better than we now do. What I have sketched is only a superficial beginning.

III. Ancient Antinomies and Modern Invariances

Because of his view of the nature of the heavens, Aristotle famously held that the world is eternal. Almost a thousand years later the Christian philosopher Philoponus launched a major attack against Aristotle on the eternity of the world. Although Christians needed to believe in the creation, not the eternity, of the world, Aquinas wisely concluded in Question 46, Part I of the *Summa Theologica* that it must be an article of faith, not demonstration, that the world had a beginning. This long and passionate debate, which I am only sketching, formed the subject of Kant's First Antinomy in the *Critique of Pure Reason.* A modern view, so I would argue, is that a decisive demonstration of the finiteness or infinitude of space and time is out of reach. It is unlikely we can ever prove the Big Bang was the absolute beginning; we certainly seem to have no clue as to how such a proof would be constructed. Note also that we even have a revivial by Hoyle and others of the view that the universe may have been through endless cycles of creation, and the Big Bang was only the beginning of the current cycle.

Ancient arguments between Aristotelians and atomists about the continuity or discreteness of matter are well-known, and were the focus of Kant's Second Antinomy. From a mathematical standpoint continuity and discreteness are radically different, and we ordinarily think of a particular feature of a mathematical model of a theory as exemplifying one or the other. On the other hand, it is difficult to conceive of an experiment that would distinguish between a physical quantity being continuous or discrete if a sufficiently small grid of discreteness is permitted. Experimental proof could only go one way: in some cases there is discreteness of a given, established coarseness. So we expect indistinguishability at a very fine level of discreteness between it and continuity. An appropriate invariance should hold, in spite of the central role of continuity in classical mathematical analysis and many other parts of mathematics. The role of continuity, so I would argue, is not as an inaccessible ontological principle, but as a powerful assumption to support the calculations, symbolic and numerical, of mathematical physics that require continuity. They work, because they are such fine approximations. If there is discreteness, it is not too coarse. Here we can take the view that a discrete and a continuous model, one each of a given set of phenomena, must satisfy theories that are mutually contradictory when restricted to the same domains of interpretation.

This same kind of contradiction arises in connection with Kant's Third Antinomy, that between deterministic and indeterministic theories, the latter having a component of absolute spontaneity, in Kant's insightful terminology. Here we have modern support of the same kind mentioned in connection with the Second Antinomy: two kinds of theories, but indistinguishable from an observational standpoint, no matter how many observations are taken, if the accuracy of the quantitative observations is bounded, i.e., not infinitely precise. Perhaps the simplest example is the ergodic one of Sinai billiards. We have a single billiard ball moving without loss of energy and satisfying the law of reflection, but with a convex reflecting obstacle in the middle of the table. The motion of the ball is ergodic. If we finitely partition the space of the table on which the ball can move and correspondingly make time discrete, starting with the simple Newtonian model, then this discrete model cannot be distinguished by observation at the discreteness level of the elements of the partition from a finite- state discrete Markov process. (The result can also be formulated in terms of flows for continuous time.)

So what I am arguing for are new invariances in old bottles, to show how enduring these philosophical and conceptual antinomies are, ready for new ideas and new interpretation on issues central to the natural sciences.

References

Suppes, P. (1993), The Transcendental Character of Determinism, *Midwest Studies in Philosophy*, Volume XVIII, pp. 242-257.

Suppes, P. (1994), Principles that Transcend Experience: Kant's Antinomies Revisited, to appear in *Proceedings of the 1994 Austrian Philosophy Conference*.

Arne Collen
Graduate School and Research Center
Saybrook Institute
San Francisco, California, USA

THE FOUNDATION OF SCIENCE

To address the question "What is the foundation of science?" I base my response on the two substantive constructs of the question: *science* and *foundation.*

Science. Science conveys to me the idea of pursuit in order to comprehend. It is first a process and second a result. The process is that of discovering and the result is the discovery. The process becomes a way of knowing and the outcome some form of knowledge. More recently, scientists have come to understand that the process is more creative than previous assumed, and consequently some forms of science involve not only discovering, but also creating. The status of knowledge has taken on a more temporary, transitory quality, as scientists create more informative and useful manifestations of knowledge, periodically revising them to better reflect their comprehension of reality.

As the phenomena of interest in my professional work concerns human beings, those sciences directly relevant to people, I term the human sciences, preoccupy my attention. As a methodologist, I concentrate my study on those research methods which scientists use to study human phenomena, and I term such methods human science research methods. As the forms of science differ, or perhaps more accurately advance, the forms of method do also.

Interestingly, scientists devote much time in the process of discovery and creation to both the phenomena under study and their methodology. By this I mean that scientists invent, refine, and improve their technologies, techniques, and various means of inquiry in their pursuit of knowledge. There is an important and reciprocal relationship between what we know and the technology we employ. Advances in science technology lead to advances in scientific knowledge, and vice versa.

However, human science is changing in another fashion. In addition to greater recognition of creation in the acts of scientists, the aim of science is undergoing a genuine expansion. This expansion is coming to redefine what we mean by science.

In traditional forms of science, it is assumed that the scientist, a skilled observer standing somewhat aloof from that which is studied, need only apply the proper methodology to reveal the workings of Nature. Answers to research questions exist; they await the clever scientist to uncover them. Moreover, in the last century, it was recognized that the knowledge of the scientist is both public and personal, and both may be socially based constructions bounded by the worldview, that is the underlying paradigm ascribed to by the scientist. One interpretation may not represent those of other scientists or general laws of Nature. Doing science extended from active reflection upon what one is doing to include interaction with the phenomenon studied and participation in an ongoing dialog and critique of findings and means of discovery. Reminiscent of the contribution of Copernicus, whose science brought about a shift from the theocentric and geocentric worldviews to the heliocentric worldview, importantly, a controversy in the middle of the 19th century divided scientists between the natural science (heliocentric) worldview leading to explanation and the human studies (anthropocentric) worldview leading to understanding. This historical development in the history of science represents the culmination of an emerging shift away from a heliocentric worldview that is still taking place today. Though the usual outcome of participation in such debates among scientists – then and now – has been to favor one position to the rejection of the other, I believe what we witness from each debate is a recognition by more scientists that multiple worldviews are relevant to science. Each one has assumptions which may be at variance from the other, and each serves somewhat different interests among scientists. More specifically, where the natural science worldview (arena 1) involves the discovery and formulation of knowledge which promotes public and consensually supported explanations of phenomena, the human–centered, humanistic worldview (arena 2) emphasizes the personal understandings of the scientists and research participants engaged in the inquiry. Each arena serves different science interests. One is neither more or less important than the other. But both exist and should be recognized and articulated.

By the middle of this century, a third arena emerged in which the main aim of the scientist became the amelioration of human conditions. This form of inquiry has become variously known as social action, social intervention, and participatory action research, and its methods of conducting science are at variance with those of the first and second arenas.

My view of the matter is that the three arenas demonstrate the multitude of interests among scientists as well as the purposes to which they apply their science. The arenas reflect the underlying paradigms that influence the conduct of inquiry. I expect more arenas to emerge with further advances of science. Therefore, in any consideration of the foundation of science, it is important to emphasis that there are different forms which science can take. These forms constitute science in its broadest pursuit of knowledge, and we need all kinds of scientists to study our complex problems.

As a methodologist, it is most challenging for me to work with the three arenas stated, because I believe that they are not contradictory or oppositional: to the contrary, they have an important complementary, often nested, interrelationship. Currently, I am witness to many scientists in Europe and the United States who are discovering that human science research methods can be combined in various ways to create a more productive, effective and informative methodology. I provide two examples. Naturalistic observation (arena 1), nonparticipant observation (arena 2) and participant observation (arena 3) may be combined to construct an observation methodology, which is often the case in ethnographic research in anthropology. In management science, a social action research project may involve a survey (arena 1), followed by interviews (arena 2), and finishing with focus group discussions (arena 3), from which the researcher converges upon the findings in order to make recommendations to improve the institution.

In summary, science consists of means of inquiry in pursuit of knowledge relevant to aims. The aims of science may be to 1) provide an explanation (publically debated theory) of the phenomenon, 2) deepen the personal understanding of the scientists and participants engaged in the inquiry, and 3) ameliorate human conditions. Of course, there are many possible aims and interests of scientists not stated here, but I have found these three to be most dominant, thus I characterize each to represent their underlying paradigm of science.

Foundation. As to the second key construct in the question, the most fundamental concepts and principles provide the foundation of science. Without them, the scientist cannot engage in doing science. Of course, it follows from the first section of this statement that this foundation consists of the methods, knowledge, and aims of scientists.

Science is first descriptive. Descriptions of phenomena require a common set of tools and a language by which scientists work. The application of fundamental concepts and principles to 1. methodology, 2. knowledge creation, utilization, and revision, and 3. scientific interest gives substance to science. Foundation means a stable dependable basis with which one can work. It also means a solid ground upon which to stand to build theory, mature personal understanding, and act in concert with others for changing living conditions.

Fundamental concepts and principles are especially those that cut across all sciences. Such exempars as observation, interpretation, triangulation, and replication are critical concepts to know, if one is to know what scientists do. Understanding these concepts is paramount in praxis of inquiry which is efficient, effective, and fruitful.

There is by no means widespread agreement among scientists as to what constitutes scientific interests, scientific method, and scientific knowledge. The very foundation of science has been challenged in regard to the assumptions scientists make about purpose, method and knowledge. Variations in position on these matters are evident in the assumptions and beliefs of scientists who work in each arena of inquiry. Is it possible that knowledge can represent explanation, understanding, and amelioration? Are the means scientists use to obtain explanation, understanding, and amelioration different forms, and legitimate forms, of scientific method? These are controversial subjects.

Remarkably, there are many constructs which unify scientists. Such a construct, for example, is information. It has an intriguing relation to knowledge. Despite variations in the definition of this term, information is an example of a basic concept that has a unifying effect on the sciences. It has become one brick, so to speak, of the foundation. It enables communication among scientists across the sciences and fosters the advancement of science. Furthermore, it is not the linguistic label itself that I emphasize, but what it stands for, that is, the phenomenon that it represents. In other words, information is isomorphic. The concepts and principles, that scientists discover which are isomorphic, are those that contribute essentially to the foundation of science. The more isomorphic the concepts and principles are, then the more generic the science.

In contrast to single constructs like information, there are interrelated sets of them that also importantly contribute to the foundation of science. General Systems Theory may be an outstanding example of a theoretical approach to science comprised of a set of interrelated constructs that tend to be isomorphic in their application across the sciences.

Methodology represents my major interest in the foundation of science. Like a stone cutter and bricklayer who must know his tools well in order to build a solid foundation for a home, a human–oriented scientist must be most familiar with various methods of human science research in order to conduct fruitful inquiry. As necessary, this activity includes development of new and innovative science technologies which advance the aims of science. At the turn of the century, advances in methodology are expanding the foundation of science. Scientists are combining methods in new ways and making every more creative the means to innovate methodology. Presently, the range of methods and possibilities to combine them are staggering. To give but one illustration: The multitude of media–related technologies today enables scientists to study human activity through not just video, email, and other electronic data trails, but additionally by means of several types of microscopic scanning into the human body and macroscopic pattern imaging from global satellite networks.

In conclusion, the foundation of science is the more enduring and substantive concepts and principles as expressed in the aims, methods, and knowledge of scientists. Although on the surface the foundation may appear static, this stasis is illusionary. Under the surface, the foundation is very turbulent. Scientists are in continuous debate over the interpretation of evidence, the proper scientific method for the phenomenon studied, and the purposes served through inquiry. Science involves reflective, critical, speculative, and creative activity. It is this activity that quarantees continued vitality and evolution of the foundation of science.

Foundations of Science
1 (1995/96), 19-39

David Freedman
Statistics Department
University of California
Berkeley, CA 94720, USA

SOME ISSUES IN THE
FOUNDATION OF STATISTICS

> "Son, no matter how far you travel, or how smart you get, always remember this: Someday, somewhere, a guy is going to show you a nice brand-new deck of cards on which the seal is never broken, and this guy is going to offer to bet you that the jack of spades will jump out of this deck and squirt cider in your ear. But, son, do not bet him, for as sure as you do you are going to get an ear full of cider."
>
> Damon Runyon[1]

Key Words: Statistics, Probability, Objectivist, Subjectivist, Bayes, de Finetti, Decision theory, Model validation, Regression.

Abstract. After sketching the conflict between objectivists and subjectivists on the foundations of statistics, this paper discusses an issue facing statisticians of both schools, namely, model validation. Statistical models originate in the study of games of chance, and have been successfully applied in the physical and life sciences. However, there are basic problems in applying the models to social phenomena; some of the difficulties will be pointed out. Hooke's law will be contrasted with regression models for salary discrimination, the latter being a fairly typical application in the social sciences.

---

[1]From 'The Idyll of Miss Sarah Brown', *Collier's Magazine*, 1933. Reprinted in *Guys and Dolls: The Stories of Damon Runyon.* Penguin Books, New York, 1992, pp.14-26. The quote is edited slightly, for continuity.

1. What is probability?

For a contemporary mathematician, probability is easy to define, as a countably additive set function on a σ-field, with a total mass of 1. This definition, perhaps cryptic for non-mathematicians, was introduced by A. N. Kolmogorov around 1930, and has been extremely convenient for mathematical work; theorems can be stated with clarity, and proved with rigor.[2]

For applied workers, the definition is less useful; countable additivity and σ-fields are not observed in nature. The issue is of a familiar type – What objects in the world correspond to probabilities? This question divides statisticians into two camps:

- the "objectivist" school, also called the "frequentists";

- the "subjectivist" school, also called the "Bayesians," after the Reverend Thomas Bayes (England, c.1701-1761).

Other positions have now largely fallen into disfavor; or example, there were "fiducial" probabilities introduced by R. A. Fisher (England, 1890-1962). Fisher was one of the two great statisticians of the century; the second, Jerzy Neyman (b. Russia, 1894; d. U.S.A. 1981), turned to objectivism after a Bayesian start. Indeed, the objectivist position now seems to be the dominant one in the field, although the subjectivists are still a strong presence. Of course, the names are imperfect descriptors. Furthermore, statisticians agree amongst themselves about as well as philosophers; many shades of opinion will be represented in each school.

2. The objectivist position

Objectivists hold that probabilities are inherent properties of the systems being studied. For a simple example, like the toss of a coin, the idea seems quite clear at first. You toss the coin, it will land heads or tails, and

---

[2]This note will give a compact statement of Kolmogorov's axioms. Let Ω be a set. By definition, a σ-field $\mathcal{F}$ is a collection of subsets of Ω, which has Ω itself as a member. Furthermore,

- $\mathcal{F}$ is closed under complementation (if $A \in \mathcal{F}$ then $A^c \in \mathcal{F}$), and
- $\mathcal{F}$ is closed under the formation of countable unions: if $A_i \in \mathcal{F}$ for $i = 1, 2, \ldots$, then $\cup_i A_i \in \mathcal{F}$.

A probability P is a non-negative, real-valued function on Ω, such that $P(\Omega) = 1$ and P is countably additive: if $A_i \in \mathcal{F}$ for $i = 1, 2, \ldots$, and the sets are pairwise disjoint, in the sense that $A_i \cap A_j = \emptyset$ for $i \neq j$, then $P(\cup_i A_i) = \sum_i P(A_i)$. A random variable X is an $\mathcal{F}$-measurable function on Ω. Informally, probabilists might say that Nature chooses $\omega \in \Omega$ according to P, and shows you $X(\omega)$; the latter would be the "observed value" of X.

the probability of heads is around 50%. A more exact value can be determined experimentally, by tossing the coin repeatedly and taking the long run relative frequency of heads. In one such experiment, John Kerrich (a South African mathematician interned by the Germans during World War II) tossed a coin 10,000 times and got 5,067 heads: the relative frequency was 5,067/10,000 = 50.67%. For an objectivist such as myself, the probability of Kerrich's coin landing heads has its own existence, separate from the data; the latter enable us to estimate the probability, or test hypothesis concerning it.

The objectivist position exposes one to certain famous difficulties. As Keynes said, "In the long run, we are all dead." Heraclitus' epigram (also out of context) is even more severe: "You can't step into the same river twice." Still, the tosses of a coin, like the throws of a die and the results of other such chance processes, do exhibit remarkable statistical regularities. These regularities can be described, predicted, analyzed by technical probability theory. Using Kolmogorov's axioms (or more primitive definitions), we can construct statistical models that correspond to empirical phenomena; although verification of the correspondence is not the easiest of tasks.

3. The subjectivist position

For the subjectivist, probabilities describe "degrees of belief." There are two camps within the subjectivist school, the "classical" and the "radical." For a "classical" subjectivist, like Bayes himself or Laplace – although such historical readings are quite tricky – there are objective "parameters" which are unknown and to be estimated from the data. (A parameter is a numerical characteristic of a statistical model for data – for instance, the probability of a coin landing heads; other examples will be given below.) Even before data collection, the classical subjectivist has information about the parameters, expressed in the form of a "prior probability distribution."

The crucial distinction between a classical subjectivist and an objectivist: the former will make probability statements about parameters – for example, in a certain coin-tossing experiment, there is a 25% chance that the probability of heads exceeds .67. However, objectivists usually do not find that such statements are meaningful; they view the probability of heads as an unknown constant, which either is – or is not – bigger than .67. In replications of the experiment, the probability of heads will always exceed .67, or never; 25% cannot be relevant. As a technical matter, if the parameter has a probability distribution given the data, it must have a "marginal" distribution – that is, a prior. On this point, objectivists and subjectivists agree; the hold-out was R. A. Fisher, whose fiducial probabilities come into existence only after data collection.

"Radical" subjectivists, like Bruno de Finetti or Jimmie Savage, differ from classical subjectivists and objectivists; radical subjectivists deny the very existence of unknown parameters. For such statisticians, probabilities express degrees of belief about observables. You pull a coin out of your pocket, and – Damon Runyon notwithstanding – they can assign a probability to the event that it will land heads when you toss it. The braver ones can even assign a probability to the event that you really will toss the coin. (These are "prior" probabilities, or "opinions.") Subjectivists can also "update" opinions in the light of the data; for example, if the coin is tossed 10 times, landing heads 6 times and tails 4 times, what is the chance that it will land heads on the 11th toss? This involves computing a "conditional" probability using Kolmogorov's calculus, which applies whether the probabilities are subjective or objective.

Here is an example with a different flavor: What is the chance that a republican will be president of the U.S. in the year 2025? For many subjectivists, this is a meaningful question, which can in principle be answered by introspection. For many objectivists, this question is beyond the scope of statistical theory. As best I can judge, however, complications will be found on both sides of the divide. Some subjectivists will not have quantifiable opinions about remote political events; likewise, there are objectivists who might develop statistical models for presidential elections, and compute probabilities on that basis.[3]

The difference between the radical and classical subjectivists rides on the distinction between parameters and observables; this distinction is made by objectivists too and is often quite helpful. (In some cases, of course, the issue may be rather subtle.) The radical subjectivist denial of parameters exposes them to certain rhetorical awkwardness; for example, they are required not to understand the idea of a tossing a coin with an unknown probability of heads. Indeed, if they admit the coin, they will soon be stuck with all the unknown parameters that were previously banished.[4]

Probability and relative frequency. In ordinary language, "probabilities"

---

[3] Models will be discussed in section 5. Those for presidential elections may not be compelling. For genetics, however, chance models are well established; and many statistical calculations are therefore on a secure footing. Much controversy remains, for example, in the area of DNA identification (*Jurimetrics*, vol. 34, no. 1, 1993).

[4] The distinction between classical and radical subjectivists made here is not often discussed in the statistical literature; the terminology is not standard. See, for instance, Diaconis and Freedman (1980a), Efron (1986), Jeffrey (1983, sec. 12.6).

are not distinguished at all sharply from empirical percentages – "relative frequencies." In statistics, the distinction may be more critical. With Kerrich's
coin, the relative frequency of heads in 10,000 tosses, 50.67%, is unlikely to
be the exact probability of heads; but it is unlikely to be very far off. For
an example with a different texture, suppose you see the following sequence
of 10 heads and 10 tails:

T H T H T H T H T H T H T H T H T H T H.

What is the probability that the next observation will be a head? In this
case, relative frequency and probability are quite different.[5]

One more illustration: United Airlines flight 140 operates daily from San
Francisco to Philadelphia. In 192 out of the last 365 days, flight 140 landed
on time. You are going to take this flight tomorrow. Is your probability of
landing on time given by 192/365? For a radical subjectivist, the question
is clear; not so for an objectivist or a classical subjectivist. Whatever the
question really means, 192/365 is the wrong answer – if you are flying on
the Friday before Christmas. This is Fisher's "relevant subset" issue; and he
seems to have been anticipated by von Mises. Of course, if you pick a day at
random from the data set, the chance of getting one with an on-time landing
is indeed 192/365; that would not be controversial. The difficulties come with
(i) extrapolation and (ii) judging the exchangeability of the data, in a useful
Bayesian phrase. Probability is a subtler idea than relative frequency.[6]

Labels do not settle the issue. Objectivists sometimes argue that they
have the advantage, because science is objective. This is not serious; "objectivist" statistical analysis must often rely on judgment and experience:

---

[5]Some readers may say to themselves that here, probability is just the relative frequency
of transitions. However, a similar but slightly more complicated example can be rigged
up for transition counts; an infinite regress lies just ahead. My point is only this: relative
frequencies are not probabilities. Of course, if circumstances are favorable, the two are
strongly connected – that is one reason why chance models are useful for applied work.

[6]To illustrate the objectivist way of handling probabilities and relative frequencies, I
consider repeated tosses of a fair coin: the probability of heads is 50%. In a sequence
of 10,000 tosses, the chance of getting between 49% and 51% heads is about 95%. In
replications of this (large) experiment, about 95% of the time, there will be between 49%
and 51% heads. On each replication, however, the probability of heads stays the same –
namely, 50%.

The strong law of large numbers provides another illustration. Consider n repeated
tosses of a fair coin. With probability 1, as $n \to \infty$, the relative frequency of heads in the
first n tosses eventually gets trapped inside the interval from 49% to 51%; ditto, for the
interval from 49.9% to 50.1%; ditto, for the interval from 49.99% to 50.01%; and so forth.
No matter what the relative frequency of heads happens to be at any given moment, the
probability of heads stays the same – namely, 50%. Probability is not relative frequency.

subjective elements come in. Likewise, subjectivists may tell you that (i) objectivists use "prior information" and (ii) are therefore closet Bayesians. Point (i) may be granted. The issue for (ii) is how prior information enters the analysis, and whether this information can be quantified or updated the way subjectivists insist it must be. The real questions are not to be settled on the basis of labels.

4. A critique of the subjectivist position

The subjectivist position seems to be internally consistent, and fairly immune to logical attack from the outside. Perhaps as a result, scholars of that school have been quite energetic in pointing out the flaws in the objectivist position. From an applied perspective, however, the subjectivist position is not free of difficulties. What are subjective degrees of belief, where do they come from, and why can they be quantified? No convincing answers have been produced. At a more practical level, a Bayesian's opinion may be of great interest to himself, and he is surely free to develop it in any way that pleases him; but why should the results carry any weight for others?

To answer the last question, Bayesians often cite theorems showing "inter-subjective agreement:" under certain circumstances, as more and more data become available, two Bayesians will come to agree: the data swamp the prior. Of course, other theorems show that the prior swamps the data, even when the size of the data set grows without bounds – particularly in complex, high-dimensional situations. (For a review, see Diaconis and Freedman, 1986.) Theorems do not settle the issue, especially for those who are not Bayesians to start with.

My own experience suggests that neither decision-makers nor their statisticians do in fact have prior probabilities. A large part of Bayesian statistics is about what you would do if you had a prior.[7] For the rest, statisticians make up priors that are mathematically convenient or attractive. Once used, priors become familiar; therefore, they come to be accepted as "natural" and are liable to be used again; such priors may eventually generate their own technical literature.

Other arguments for the Bayesian position. Coherence. There are well-

---

[7] Similarly, a large part of objectivist statistics is about what you would do if you had a model; and all of us spend enormous amounts of energy finding out what would happen if the data kept pouring in. I wish we could learn to look at the data more directly, without the fictional models and priors. On the same wish-list: we stop pretending to fix bad designs and inadequate measurements by modeling.

known theorems, including (Freedman and Purves, 1969), showing that stubborn non-Bayesian behavior has costs. They can make a "dutch book," and extract your last penny – if you are generous enough to cover all the bets needed to prove the results.[8] However, most of us don't bet at all; even the professionals bet on relatively few events. Thus, coherence has little practical relevance. (Its rhetorical power is undeniable – who wants to be incoherent?)

Rationality. It is often urged that to be rational is to be Bayesian. Indeed, there are elaborate axiom systems about preference orderings, acts, consequences, and states of nature, whose conclusion is – that you are a Bayesian. The empirical evidence shows, fairly clearly, that those axioms do not describe human behavior at all well. The theory is not descriptive; people do not have stable, coherent prior probabilities.

Now the argument shifts to the "normative:" if you were rational, you would obey the axioms, and be a Bayesian. This, however, assumes what must be proved. Why would a rational person obey those axioms? The axioms represent decision problems in schematic and highly stylized ways. Therefore, as I see it, the theory addresses only limited aspects of rationality. Some Bayesians have tried to win this argument on the cheap: to be rational is, by definition, to obey their axioms. (Objectivists do not always stay on the rhetorical high road either.)

Detailed examination of the flaws in the normative argument is a complicated task, beyond the scope of the present article. In brief, my position is this. Many of the axioms, on their own, have considerable normative force. For example, if I am found to be in violation of the "sure thing principle," I would probably reconsider.[9] On the other hand, taken as a whole, decision theory seems to have about the same connection to real decisions as war games played on a table do to real wars.

What are the main complications? For some events, I may have a rough idea of likelihood: one event is very likely, another is unlikely, a third is uncertain. However, I may not be able to quantify these likelihoods, even to one or two decimal places; and there will be many events whose probabilities are simply unknown – even if definable.[10] Likewise, there are some benefits that can be assessed with reasonable accuracy; others can be estimated only

---

[8] A "dutch book" is a collection of bets on various events such that the bettor makes money, no matter what the outcome.

[9] According to the "sure thing principle," if I prefer *A* to *B* given that *C* occurs, and I also prefer *A* to *B* given that *C* does not occur, I must prefer *A* to *B* when I am in doubt as to the occurrence of *C*.

[10] Although one-sentence concessions in a book are not binding, Savage (1954, p.59) does say that his theory "is a code of consistency for the person applying it, not a system of

to rough orders of magnitude; in some cases, quantification may not be possible at all. Thus, utilities may be just as problematic as priors.

The theorems that derive probabilities and utilities from axioms push the difficulties back one step.[11] In real examples, the existence of many states of nature must remain unsuspected. Only some acts can be contemplated; others are not imaginable until the moment of truth arrives. Of the acts that can be imagined, the decision-maker will have preferences between some pairs but not others. Too, common knowledge suggests that consequences are often quite different in the foreseeing and in the experiencing.

Intransitivity would be an argument for revision, although not a decisive one; for example, a person choosing among several job offers might well have intransitive preferences, which it would be a mistake to ignore. By way of contrast, an arbitrageur who trades bonds intransitively is likely to lose a lot of money. (There is an active market in bonds, while the market in job offers – largely non-transferable – must be rather thin; the practical details make a difference.) The axioms do not capture the texture of real decision making. Therefore, the theory has little normative force.

The fallback defense. Some Bayesians will concede much of what I have

---

predictions about the world"; and personal probabilities can be known "only roughly."

Another comment on this book may be in order. According to Savage (1954, pp.61-62), "on no ordinary objectivistic view would it be meaningful, let alone true, to say that on the basis of the available evidence it is very improbable, though not impossible, that France will become a monarchy within the next decade." As anthropology of science, this seems wrong. I make qualitative statements about likelihoods and possibilities, and expect to be understood; I find such statements meaningful when others make them. Only the quantification seems problematic: What would it mean to say that P(France will become a monarchy) = .0032? Many objectivists of my acquaintance share such views; although caution is in order when extrapolating from such a sample of convenience.

[11]The argument in the text is addressed to readers who have some familiarity with the axioms. This note gives a very brief review; Kreps (1988) has a chatty and sympathetic discussion (although some of the details are not quite in focus); Le Cam (1977) is more technical and critical.

In the axiomatic setup, there is a space of "states of nature," like the possible orders in which horses finish a race. There is another space of "consequences"; these can be pecuniary or non-pecuniary (win $1,000, lose $5,000, win a weekend in Philadelphia, etc.). Mathematically, an "act" is a function whose domain is the space of states of nature, and whose values are consequences. You have to choose an act: that is the decision problem. Informally, if you choose the act f, and the state of nature happens to be s, you enjoy (or suffer) the consequence $f(s)$. For example, if you bet on those horses, the payoff depends on the order in which they finish: the bet is an act, and the consequence depends on the state of nature. The set of possible states of nature, the set of possible consequences, and the set of possible acts are all viewed as fixed and known. You are supposed to have a transitive preference ordering on the acts, not just the consequences. The sure thing principle is an axiom in Savage's setup.

said: the axioms are not binding; rational decision-makers may have neither priors nor utilities. Still, the following sorts of arguments can be heard. The decision-maker must have some ideas about relative likelihoods for a few events; a prior probability can be made up to capture such intuitions, at least in gross outline. The details (for instance, that distributions are normal) can be chosen on the basis of convenience. A utility function can be put together using similar logic: the decision-maker must perceive some consequences as very good, and big utility numbers can be assigned to these; he must perceive some other consequences as trivial, and small utilities can be assigned to those; in between is in between. The Bayesian engine can now be put to work, using such approximate priors and utilities. Even with these fairly crude approximations, Bayesian analysis is held to dominate other forms of inference: that is the fallback defense.

Here is my reaction to such arguments. Approximate Bayesian analysis may in principle be useful. That this mode of analysis dominates other forms of inference, however, seems quite debatable. In a statistical decision problem, where the model and loss function are given, Bayes procedures are often hard to beat, as are objectivist likelihood procedures; with many of the familiar textbook models, objectivist and subjectivist procedures should give similar results if the data set is large. There are sharp mathematical theorems to back up such statements.[12] On the other hand, in real problems

---

[12] Wald's idea of a statistical decision problem can be sketched, as follows. There is an unobservable parameter. Corresponding to each parameter value θ, there is a known probability distribution P_θ for an observable random quantity X. (This family of probability distributions is a "statistical model" for X, with parameter θ.) There is a set of possible "decisions"; there is a "loss function" $L(d, \theta)$ which tells you how much is lost by making the decision d when the parameter is really θ. (For example, d might be an estimate of θ, and loss might be squared error.) You have to choose a "decision rule," which is a mapping from observed values of X to decisions. Your objective is to minimize "risk," that is, expected loss.

A comparison with the setup in note 11 may be useful. The "state of nature" seems to consist of the observable value of X, together with the unobservable value θ of the parameter. The "consequences" are the decisions, and "acts" are decision rules. (The conflict in terminology is regrettable, but there is no going back.) The utility function is replaced by L, which is given but depends on θ as well as d.

A Bayes' procedure is optimal in that its risk cannot be reduced for all values of θ; any such "admissible" procedure is a limit of Bayes' procedures ("the complete class theorem"). The maximum likelihood estimator is "efficient"; and its sampling distribution is close to the posterior distribution of θ by the "Bernstein-von Mises theorem," which is actually due to Laplace. More or less stringent regularity conditions must be imposed to prove any of these results, and some of the theorems must be read rather literally; Stein's paradox and Bahadur's example should at least be mentioned.

Standard monographs and texts include Berger (1985), Berger and Wolpert (1988), Bickel and Doksum (1977), Casella and Berger (1990), Ferguson (1967), Le Cam (1986),

– where models and loss functions are mere approximations – the optimality of Bayes procedures cannot be a mathematical proposition. And empirical proof is conspicuously absent.

If we could quantify breakdowns in model assumptions, or degrees of error in approximate priors and loss functions, the balance of argument might shift considerably. The rhetoric of "robustness" may suggest that such error analyses are routine. This is hardly the case even for the models. For priors and utilities, the position is even worse, since the entities being approximated do not have any independent existence – outside the Bayesian framework that has been imposed on the problem.

de Finetti's theorem. Suppose you are a radical subjectivist, watching a sequence of 0's and 1's. In your prior opinion, this sequence is exchangeable: permuting the order of the variables will not change your opinion about them. A beautiful theorem of de Finetti's asserts that your opinion can be represented as coin tossing, the probability of heads being selected at random from a suitable prior distribution. This theorem is often said to "explain" subjective or objective probabilities, or justify one system in terms of the other.[13]

Such claims cannot be right. What the theorem does is this: it enables the subjectivist to discover features of his prior by mathematical proof, rather than introspection. For example, suppose you have an exchangeable prior about those 0's and 1's. Before data collection starts, de Finetti will prove to you by pure mathematics that in your own opinion the relative frequency of 1's among the first n observations will almost surely converge to a limit as $n \to \infty$. (Of course, the theorem has other consequences too,

---

Lehmann (1983, 1986), and Rao (1973). The Bernstein-von Mises theorem is discussed in Le Cam and Yang (1990) and Prakasa Rao (1987).

Of course, in many contexts, Bayes procedures and frequentist procedures will go in opposite directions; for a review, see Diaconis and Freedman (1986). These references are all fairly technical.

[13] Diaconis and Freedman (1980ab, 1981) review the issues and the mathematics. The first-cited paper is relatively informal; the second gives a version of de Finetti's theorem applicable to a finite number of observations, with bounds; the last gives a fairly general mathematical treatment of partial exchangeability, with numerous examples and it is quite technical. More recent work is described in Diaconis and Freedman (1988, 1990).

The usual hyperbole can be sampled in Kreps (1988, p.145): de Finetti's theorem is "the fundamental theorem of statistical inference – the theorem that from a subjectivist point of view makes sense out of most statistical procedures." This interpretation of the theorem fails to distinguish between what is assumed and what is proved. It is the assumption of exchangeability that enables you to predict the future from the past, at least to your own satisfaction, not the conclusions of the theorem or the elegance of the proof. If have an exchangeable prior, the statistical world looks like your oyster, de Finetti or no de Finetti.

but all have the same logical texture.)

This notion of "almost surely," and the limiting relative frequency, are features of your opinion not of any external reality. ("Almost surely" means with probability 1, and the probability in question is your prior.) Indeed, if you had not noticed these consequences of your prior by introspection, and now do not like them, you are free to revise your opinion – which will have no impact outside your head. What the theorem does is to show how various aspects of your prior opinion are related to each other. That is all the theorem can do, because the conditions of the theorem are conditions on the prior alone.

To illustrate the difficulty, I cite an old friend rather than a new enemy. According to Jeffrey (1983, p.199), de Finetti's result proves "your subjective probability measure [is] a certain mixture or weighted average of the various possible objective probability measures" – an unusually clear statement of the interpretation that I deny. Each of Jeffrey's "objective" probability measures governs the tosses of a p-coin, where p is your limiting relative frequency of 1's. (Of course, p has a probability distribution of its own, in your opinion.) Thus, p is a feature of your opinion, not of the real world: the mixands in de Finetti's theorem are "objective" only by terminological courtesy. In short, the "p-coins" that come out of de Finetti's theorem are just as subjective as the prior that went in.

To sum up. The theory – as developed by Ramsey, von Neumann and Morgenstern, de Finetti, and Savage, among others – is great work. They solved an important historical problem, of interest to economists, mathematicians, statisticians, and philosophers alike. On a more practical level, the language of subjective probability is evocative; some investigators find the consistency of Bayesian statistics to be a useful discipline; for some (including me), the Bayesian approach can suggest statistical procedures whose behavior is worth investigating. But the theory is not a complete account of rationality, or even close. Nor is it the prescribed solution for any large number of problems in applied statistics, at least as I see matters.

5. Statistical models

Of course, statistical models are applied not only to coin tossing but also to more complex systems. For example, "regression models" are widely used in the social sciences, as indicated below; such applications raise serious epistemological questions. (This idea will be developed from an objectivist perspective, but similar issues are felt in the other camp.)

The problem is not purely academic. The census suffers an undercount, more severe in some places than others; if certain statistical models are to be

believed, the undercount can be corrected – moving seats in Congress and
millions of dollars a year in entitlement funds (*Survey Methodology*, vol. 18,
no. 1, 1992; *Jurimetrics*, vol. 34, no. 1, 1993; *Statistical Science*, vol. 9, no.
4, 1994). If yet other statistical models are to be believed, the veil of secrecy
can be lifted from the ballot box, enabling the experts to determine how
racial or ethnic groups have voted – a crucial step in litigation to enforce
minority voting rights (*Evaluation Review*, vol. 15, no. 6, 1991; Klein and
Freedman, 1993).

Here, I begin with a (relatively) non-controversial example from physics –
Hooke's law: strain is proportional to stress. (This law is named after Robert
Hooke, England, 1653-1703.) We will have some number n of observations.
For the ith observation, indicated by the subscript i, we hang weight$_i$ on
a spring. The length of the spring is measured as length$_i$. The regression
model says that (for quite a large range of weights [14]),

$$\text{length}_i = a + b \times \text{weight}_i + \varepsilon_i. \tag{1}$$

The "error" term ε_i is needed because measured length will not be exactly
equal to $a + b \times$ weight. If nothing else, measurement error must be reckoned
with. We model ε_i as a sequence of draws, made at random with replacement
from a box of tickets; each ticket shows a potential error – the ε_i that will be
realized if that ticket is the ith one drawn. The average of all the potential
errors in the box is assumed to be 0. In more standard terminology, the ε_i
are assumed to be "independent and identically distributed with mean 0."
Such assumptions can present difficult scientific issues, because error terms
are not observable.

In equation (1), a and b are parameters, unknown constants of nature
that characterize the spring: a is the length of the spring under no load, and
b is stretchiness – the increase in length per unit increase in weight. These
parameters are not observable, but they can be estimated by "the method
of least squares," developed by Adrien-Marie Legendre (France, 1752-1833)
and Carl Friedrich Gauss (Germany, 1777-1855) to fit astronomical orbits.
Basically, you choose the values of $\hat{a}$ and $\hat{b}$ to minimize the sum of the
squared "prediction errors," $\sum_i e_i^2$, where e_i is the prediction error for the
ith observation:[15]

---

[14]With large-enough weights, a quadratic term will be needed in equation (1). Moreover,
beyond some point, the spring passes its "elastic limit" and snaps.

[15]The residual e_i is observable, but is only an approximation to the disturbance term ε_i
in (1); that is because the estimates $\hat{a}$ and $\hat{b}$ are only approximations to the parameters a
and b.

$$e_i = \text{length}_i - \hat{a} - \hat{b} \times \text{weight}_i. \tag{2}$$

These prediction errors are often called "residuals:" they measure the difference between the actual length and the predicted length, the latter being $\hat{a} + \hat{b} \times \text{weight}$.

No one really imagines there to be a box of tickets hidden in the spring. However, the variability of physical measurements (under many but by no means all circumstances) does seem to be remarkably like the variability in draws from a box. This is Gauss' model for measurement error. In short, statistical models can be constructed that correspond rather closely to empirical phenomena.

I turn now to social-science applications. A case study would take us too far afield, but a stylized example – regression analysis used to demonstrate sex discrimination in salaries, adapted from (Kaye and Freedman, 1994) – may give the idea. We use a regression model to predict salaries (dollars per year) of employees in a firm from:

- education (years of schooling completed),

- experience (years with the firm),

- the dummy variable "man," which takes the value 1 for men and 0 for women.

Employees are indexed by the subscript i; for example, salary_i is the salary of the ith employee.

The equation is [16]

$$\text{salary}_i = a + b \times \text{education}_i + c \times \text{experience}_i + d \times \text{man}_i + \varepsilon_i. \tag{3}$$

Equation (3) is a statistical model for the data, with unknown parameters a, b, c, d; here, a is the "intercept" and the others are "regression coefficients"; ε_i is an unobservable error term. This is a formal analog of Hooke's law (1); the same assumptions are made about the errors. In other words, an employee's salary is determined as if by computing

$$a + b \times \text{education} + c \times \text{experience} + d \times \text{man}, \tag{4}$$

---

[16]Such equations are suggested, somewhat loosely, by "human capital theory." However, there remains considerable uncertainty about which variables to put into the equation, what functional form to assume, and how error terms are supposed to behave. Adding more variables is no panacea: Freedman (1983), Clogg and Haritou (1994).

then adding an error drawn at random from a box of tickets. The display
(4) is the expected value for salary given the explanatory variables (education, experience, man); the error term in (3) represents deviations from the
expected.

The parameters in (3) are estimated from the data using least squares. If
the estimated coefficient d for the dummy variable turns out to be positive
and "statistically significant" (by a "t-test"), that would be taken as evidence of disparate impact: men earn more than women, even after adjusting
for differences in background factors that might affect productivity. Education and experience are entered into equation (3) as "statistical controls,"
precisely in order to claim that adjustment has been made for differences in
backgrounds.

Suppose the estimated equation turns out as follows:

$$\text{predicted salary} = \$7,100 + \$1,300 \times \text{education} +$$

$$\$2,200 \times \text{experience} + \$700 \times \text{man}. \tag{5}$$

That is, $\hat{a} = \$7,100$, $\hat{b} = \$1,300$, and so forth. According to equation (5),
every extra year of education is worth on average \$1,300; similarly, every
extra year of experience is worth on average \$2,200; and, most important,
men get an premium of \$700 over women with the same education and
experience, on average.

A numerical example will illustrate (5). A male employee with 12 years
of education (high school) and 10 years of experience would have a predicted
salary of

$$\$7,100 + \$1,300 \times 12 + \$2,200 \times 10 + \$700 \times 1 =$$

$$\$7,100 + \$15,600 + \$22,000 + \$700 = \$45,400. \tag{6}$$

A similarly situated female employee has a predicted salary of only

$$\$7,100 + \$1,300 \times 12 + \$2,200 \times 10 + \$700 \times 0 =$$

$$\$7,100 + \$15,600 + \$22,000 + \$0 = \$44,700. \tag{7}$$

Notice the impact of the dummy variable: \$700 is added to (6), but not to
(7).

A major step in the argument is establishing that the estimated coefficient of the dummy variable in (3) is "statistically significant." This step

turns out to depend on the statistical assumptions built into the model. For instance, each extra year of education is assumed to be worth the same (on average) across all levels of experience, both for men and women. Similarly, each extra year of experience is worth the same across all levels of education, both for men and women. Furthermore, the premium paid to men does not depend systematically on education or experience. Ability, quality of education, or quality of experience are assumed not to make any systematic difference to the predictions of the model.

The story about the error term – that the ε's are independent and identically distributed from person to person in the data set – turns out to be critical for computing statistical significance. Discrimination cannot be proved by regression modeling unless statistical significance can be established, and statistical significance cannot be established unless conventional presuppositions are made about unobservable error terms.

Lurking behind the typical regression model will be found a host of such assumptions; without them, legitimate inferences cannot be drawn from the model. There are statistical procedures for testing some of these assumptions. However, the tests often lack the power to detect substantial failures. Furthermore, model testing may become circular; breakdowns in assumptions are detected, and the model is redefined to accommodate. In short, hiding the problems can become a major goal of model building.

Using models to make predictions of the future, or the results of interventions, would be a valuable corrective. Testing the model on a variety of data sets – rather than fitting refinements over and over again to the same data set – might be a good second-best (Ehrenberg and Bound, 1993). With Hooke's law (1), the model makes predictions that are relatively easy to test experimentally. For the salary discrimination model (3), validation seems much more difficu:. Thus, built into the equation is a model for non-discriminatory behavior: the coefficient d vanishes. If the company discriminates, that part of the model cannot be validated at all.

Regression models like (3) are widely used by social scientists to make causal inferences; such models are now almost a routine way of demonstrating counter-factuals. However, the "demonstrations" generally turn out to be depend on a series of untested, even unarticulated, technical assumptions. Under the circumstances, reliance on model outputs may be quite unjustified. Making the ideas of validation somewhat more precise is a serious problem in the philosophy of science. That models should correspond to reality is, after all, a useful but not totally straightforward idea – with some history to it. Developing models, and testing their connection to the phenomena, is a serious problem in statistics.[17]

Standard errors, t-statistics, and statistical significance. The "standard error" of $\hat{d}$ measures the likely difference between $\hat{d}$ and d, due to the action of the error terms in equation (3). The "t-statistic" is $\hat{d}$ divided by its standard error. Under the "null hypothesis" that $d = 0$, there is only about a 5% chance that $|t| > 2$. Such a large value of t would demonstrate "statistical significance." Of course, the parameter d is only a construct in a model. If the model is wrong, the standard error, t-statistic, and significance level are rather difficult to interpret.

Even if the model is granted, there is a further issue: the 5% is a probability for the data given the model, namely, $P\{|t| > 2 \parallel d = 0\}$. However, the 5% is often misinterpreted as $P\{d = 0 \mid \text{data}\}$. Indeed, this misinterpretation is a commonplace in the social-science literature, and seems to have been picked up by the courts from expert testimony.[18] For an objectivist, $P\{d = 0 \mid \text{data}\}$ makes no sense: parameters do not exhibit chance variation. For a subjectivist, $P\{d = 0 \mid \text{data}\}$ makes good sense, but its computation via the t-test is grossly wrong, because the prior probability that $d = 0$ has not been taken into account: the calculation exemplifies the "base rate fallacy." Power matters too.

(The single vertical bar "$\mid$" is standard notation for conditional proba-

---

[17] For more discussion in the context of real examples, with citations to the literature of model validation, see Freedman (1985, 1987, 1991, 1994). Many recent issues of *Sociological Methodology* have essays on this topic. Also see Oakes (1990), who discusses modeling issues, significance tests, and the objectivist-subjectivist divide.

[18] Some legal citations may be of interest (Kaye and Freedman, 1994): Waisome v. Port Authority, 948 F.2d 1370, 1376 (2d Cir. 1991) ("Social scientists consider a finding of two standard deviations significant, meaning there is about 1 chance in 20 that the explanation for a deviation could be random"); Rivera v. City of Wichita Falls, 665 F.2d 531, 545 n.22 (5th Cir. 1982) ("A variation of two standard deviations would indicate that the probability of the observed outcome occurring purely by chance would be approximately five out of 100; that is, it could be said with a 95% certainty that the outcome was not merely a fluke."); Vuyanich v. Republic Nat'l Bank, 505 F. Supp. 224, 271 (N.D. Tex. 1980), vacated and remanded, 723 F.2d 1195 (5th Cir. 1984) ("if a 5% level of significance is used, a sufficiently large t-statistic for the coefficient indicates that the chances are less than one in 20 that the true coefficient is actually zero.").

An example from the underlying technical literature may also be of interest. According to (Fisher, 1980, p.717), "in large samples, a t-statistic of approximately two means that the chances are less than one in twenty that the true coefficient is actually zero and that we are observing a larger coefficient just by chance ... A t-statistic of approximately two and one half means the chances are only one in one hundred that the true coefficient is zero ..." No. If the true coefficient is zero, there is only one chance in one hundred that $|t| > 2.5$. (Frank Fisher is a well known econometrician who often testifies as an expert witness, although I do not believe he figures in any of the cases cited above.)

bility. The double vertical bar "∥" is not standard; Bayesians might want to read this as a conditional probability; for an objectivist, ∥ is intended to mean "computed on the assumption that ...")

Statistical models and the problem of induction. How do we learn from experience? What makes us think that the future will be like the past? With contemporary modeling techniques, such questions are easily answered – in form if not in substance.

- The objectivist invents a regression model for the data, and assumes the error terms to be independent and identically distributed; "iid" is the conventional abbreviation. It is this assumption of iid-ness that enables us to predict data we have not seen from a training sample – without doing the hard work of validating the model.

- The classical subjectivist invents a regression model for the data, assumes iid errors, and then makes up a prior for unknown parameters.

- The radical subjectivist adopts an exchangeable or partially exchangeable prior, and calls you irrational or incoherent (or both) for not following suit.

In our days, serious arguments have been made from data. Beautiful, delicate theorems have been proved; although the connection with data analysis often remains to be established. And an enormous amount of fiction has been produced, masquerading as rigorous science.

6. Conclusions

I have sketched two main positions in contemporary statistics, objectivist and subjectivist, and tried to indicate the difficulties. Some questions confront statisticians from both camps: How do statistical models connect with reality? What areas lend themselves to investigation by statistical modeling? When are such investigations likely to be sterile?

These questions have philosophical components as well as technical ones. I believe model validation to be a central issue. Of course, many of my colleagues will be found to disagree. For them, fitting models to data, computing standard errors, and performing significance tests is "informative," even though the basic statistical assumptions (linearity, independence of errors, etc.) cannot be validated. This position seems indefensible, nor are the consequences trivial. Perhaps it is time to reconsider.

Acknowledgments. I would like to thank Dick Berk, Cliff Clogg, Persi Diaconis, Joe Eaton, Neil Henry, Paul Humphreys, Lucien Le Cam, Diana Petitti, Brian Skyrms, Terry Speed, Steve Turner, Amos Tversky, Ken Wachter and Don Ylvisaker for many helpful suggestions – some of which I could implement.

References

Bayes, Thomas (1764), An Essay Towards Solving a Problem in the Doctrine of Chances, *Philosophical Transactions of the Royal Society of London* **53**, 370-418.

Berger, J. (1985), *Statistical Decision Theory and Bayesian Analysis.* 2nd ed. New York, Springer-Verlag.

Berger, J. and Wolpert, R. (1988), *The Likelihood Principle.* 2nd ed. Hayward, Calif., Institute of Mathematical Statistics.

Bickel, P. J. and Doksum, K. A. (1977), *Mathematical Statistics: Basic Ideas and Selected Topics.* San Francisco, Holden-Day.

Box, G. E. P. and Tiao, G. C. (1992), *Bayesian Inference in Statistical Analysis.* New York, Wiley.

Casella, G. and Berger, R. L. (1990), *Statistical Inference.* Pacific Grove, Calif., Wadsworth & Brooks/Cole.

Clogg, C. C. and Haritou, A. (1994), The Regression Method of Causal Inference and a Dilemma with this Method. Technical report, Department of Sociology, Pennsylvania State University. To appear in V. McKim and S. Turner (eds), *Proceedings of the Notre Dame Conference on Causality in Crisis.*

de Finetti, B. (1959), *La probabilità, la statistica, nei rapporti con l'induzione, secondo diversi punti di vista.* Rome, Centro Internazionale Matematica Estivo Cremonese. English translation in de Finetti (1972).

de Finetti, B. (1972), *Probability, Induction, and Statistics.* New York, Wiley.

Diaconis, P. and Freedman, D. (1980a), de Finetti's Generalizations of Exchangeability, pp.233-50 in Richard C. Jeffrey (ed), *Studies in Inductive Logic and Probability.* Vol. 2. Berkeley, University of California Press.

Diaconis, P. and Freedman, D. (1980b), Finite Exchangeable Sequences, *Annals of Probability* **8**, 745-64.

Diaconis, P. and Freedman, D. (1981), Partial Exchangeability and Sufficiency, pp.205-36 in *Proceedings of the Indian Statistical Institute Golden Jubilee International Conference on Statistics: Applications and New Directions. Sankhya.* Calcutta, Indian Statistical Institute.

Diaconis, P. and Freedman. D. (1986), On the Consistency of Bayes' Estimates, *Annals of Statistics* 14, 1-87, with discussion.

Diaconis, P. and Freedman. D. (1988), Conditional Limit Theorems for Exponential Families and Finite Versions of de Finetti's Theorem, *Journal of Theoretical Probability* 1, 381-410.

Diaconis, P. and Freedman. D. (1990), Cauchy's Equation and de Finetti's Theorem, *Scandinavian Journal of Statistics* 17, 235-50.

Efron, B. (1986), Why Isn't Everyone a Bayesian? *The American Statistician* 40, 1-11, with discussion.

Ehrenberg, A. S. C. and Bound, J. A. (1993), Predictability and Prediction, *Journal of the Royal Statistical Society,* **Series A, Part 2, 156,** 167-206.

Ferguson, T. (1967), *Mathematical Statistics: a Decision Theoretic Approach.* New York, Academic Press.

Fisher, F. M. (1980), Multiple Regression in Legal Proceedings, *Columbia Law Review* 80, 702-36.

Fisher, R. A. (1959), *Smoking: The Cancer Controversy.* Edinburgh, Oliver & Boyd. see pp.25-29 on relevant subsets.

Freedman, D. (1983), A Note on Screening Regression Equations, *The American Statistician* 37, 152-55.

Freedman, D. (1985), Statistics and the Scientific Method, pp.343-90 in W. M. Mason and S. E. Fienberg (eds), *Cohort Analysis in Social Research: Beyond the Identification Problem.* New York, Springer.

Freedman, D. (1987), As Others See Us: A Case Study in Path Analysis, *Journal of Educational Statistics* 12, no. 2, 101-223, with discussion.

Freedman, D. (1991), Statistical Models and Shoe Leather, chapter 10 in Peter Marsden (ed), *Sociological Methodology 1991*, with discussion.

Freedman, D. (1994), From Association to Causation Via Regression. Technical report no. 408, Statistics Department, University of California, Berkeley. To appear in V. McKim and S. Turner (eds), *Proceedings of the Notre Dame Conference on Causality in Crisis.*

Freedman, D. and Purves, R. (1969), Bayes Method for Bookies, *Annals of Mathematical Statistics* 40, 1177-86.

Freedman, D., Pisani, R., Purves, R. and Adhikari, A. (1991), *Statistics,* 2nd ed. New York, Norton.

Gatsonis, C. et al., eds. (1993), *Case Studies in Bayesian Statistics.* New York, Springer-Verlag, Lecture Notes in Statistics, vol. 83.

Gauss, C. F. (1809), *Theoria Motus Corporum Coelestium.* Hamburg, Perthes et Besser. Reprinted in 1963 by Dover, New York.

Jeffrey, R. C. (1983), *The Logic of Decision.* 2nd ed. University of Chicago

Press.

Kahneman, D., Slovic, P. and Tversky, A., eds. (1982), *Judgment under Uncertainty: Heuristics and Biases.* Cambridge University Press.

Kaye, D. and Freedman, D. (1994), *Reference Manual on Statistics.* Washington, D.C., Federal Judicial Center.

Klein, S. and Freedman, D. (1993), Ecological Regression in Voting Rights Cases, *Chance Magazine* **6**, 38-43.

Kolmogorov, A. N. (1933), Grundbegriffe der Wahrscheinlichkeitstheorie, *Ergebnisse Mathematische* **2** no. 3.

Kreps, D. (1988), *Notes on the Theory of Choice.* Boulder, Westview Press.

Laplace, P. S. (1774), Memoire sur la probabilité des causes par les évenements, *Memoires de mathématique et de physique presentés a l'académie royale des sciences, par divers savants, et lûs dans ses assemblées* **6**. Reprinted in Laplace's *Oeuvres Complètes* **8**, 27-65. English translation by S. Stigler (1986), *Statistical Science* **1**, 359-378.

Le Cam, Lucien M. (1977), A Note on Metastatistics or "An Essay Toward Stating a Problem in the Doctrine of Chances," *Synthese* **36**, 133-60.

Le Cam, Lucien M. (1986), *Asymptotic Methods in Statistical Decision Theory.* New York, Springer-Verlag.

Le Cam, Lucien M. and Yang, Grace Lo (1990), *Asymptotics in Statistics: Some Basic Concepts.* New York, Springer-Verlag.

Lehmann, E. (1986), *Testing Statistical Hypotheses.* 2nd ed. Pacific Grove, Calif., Wadsworth & Brooks/Cole.

Lehmann, E. (1983), *Theory of Point Estimation.* Pacific Grove, Calif., Wadsworth & Brooks/Cole.

McNeil, B., Pauker, S., Sox, H. Jr., and Tversky, A. (1982), On the Elicitation of Preferences for Alternative Therapies, *New England Journal of Medicine* **306**, 1259-62.

Oakes, M. (1990), *Statistical Inference.* Chestnut Hill, Mass., Epidemiology Rescurces, Inc.

O'Hagan, A. (1988), *Probability: Methods and Measurement.* London, Chapman and Hall.

Peirce, C. S. (1878), The Doctrine of Chances, *Popular Science Monthly* **12** (March 1878), pp.604-615. Reprinted as pp.142-54 in N. Houser and C. Kloesel (eds) (1992), *The Essential Peirce.* Indiana University Press.

Popper, K. (1983), *Realism and the Aim of Science.* Totowa, N.J., Rowman and Littlefield.

Prakasa Rao, B. L. S. (1987), *Asymptotic Theory of Statistical Inference.* New York, Wiley.

Ramsey, F. P. (1926), in R. B. Braithwaite (1931), *The Foundations of Mathematics and other Logical Essays.* London, Routledge and Kegan Paul.

Rao, C. R. (1973), *Linear Statistical Inference and Its Applications.* 2nd ed. New York, Wiley.

Savage, Leonard J. (1972), *The Foundations of Statistics.* 2d rev. ed. New York, Dover Publications.

Stigler, S. (1986), *The History of Statistics.* Harvard University Press.

Tversky, A. and Kahneman, D. (1986), Rational Choice and the Framing of Decisions, *Journal of Business* **59**, no. 4, part 2, pp.S251-78.

Tversky, A. and Kahneman, D. (1983), Extensional versus Intuitive Reasoning: The Conjunction Fallacy in Probability Judgment, *Psychological Review* **90**, 293-315.

von Mises, R. (1964), *Mathematical Theory of Probability and Statistics.* H. Geiringer (ed). New York, Academic Press.

von Neumann, J. and Morgenstern, O. (1944), *Theory of Games and Economic Behavior.* Princeton University Press.

Foundations of Science
1 (1995/96), 41-67

COMMENTS ON DAVID FREEDMAN'S PAPER

James Berger
Statistics Department
Purdue University
West Lafayette
IN 47907 - 1399, USA

DISCUSSION OF DAVID FREEDMAN'S "SOME ISSUES IN THE FOUNDATIONS OF STATISTICS"

Key Words: Objective Bayesians, Significance testing, Interpreting P - values.

Abstract. While results from statistical modelling too often receive blind acceptance, we question whether there is any real alternative to use of modelling. This does not diminish the main point of Professor Freedman, which is that healthy scepticism towards models is needed. While agreeing with many of Professor Freedman's points concerning the "objectivist" debate, we argue that there is a Bayesian school of objectivists that possesses considerable advantages over the classical objectivist school. At the least, the "debate" needs to be enlarged to include this school.

Introduction. Professor Freedman has written a very stimulating but provocative paper. I find myself agreeing with almost all of the specific arguments that he makes, and yet disagreeing with some of the conclusions. This disagreement hinges on arguments that are not raised or extensively discussed in the paper, and which I will briefly address. Since statistical modelling is perhaps the most important issue, I will begin there.

Statistical Models. Nobody wants statistical modelling to be misused, and most statistitians and scientists can point to instances of severe misuse. And Professor Freedman is undoubtedly correct that much more credibility is placed on model-based statistical analysis than it typically deserves. But what are we to do? Going back to the "good old days " when decisions were made by guesses based on crude data summaries is not a real option. For one thing it is now sociologically impossible to remove statistical analysis from domains into which it has penetrated. Furthermore, it is hard to argue that those really were the good old days. I would hate to see medical policy decision-making return to the days when it was based on anecdotal cases and uncontrolled experiments. My personal experiences also do not lead me to reject formal statistical modelling. For instance, through recent involvement with policy issues related to fuel efficiency of cars, I discovered that policy was being proposed based on highly suspicious expert judgement, suspicious in that they ignored even the most obvious issues such as high correlations among relevant factors. I am sure that Professor Freedman could find many arguable assumptions in our own model-based modelling of the problem, but am confident that the answers so obtained were much more reasonable than the previous answers.

There is a sense in which Professor Freedman is properly taking us all to task. There seems to be a strong tendency among statistical users (at all levels) to use the latest and most elaborate statistical techniques, without proper understanding of the limitations of the techniques or their implications. I, of course, am raising this particular point of agreement to launch into an argument: one of the strong motivations for Bayesian analysis is that it results in much more understandable analyses; the assumptions (the prior probabilities) are clear, and the conclusions are easily interpretable.

As an example of the latter, consider standard significance testing, as briefly outlined by Professor Freedman. He observes that there is possibly a great difference between the classical significance level and the probability that the null hypothesis is true, given the data. To elaborate on this point, suppose that one were testing a long series of drugs for effectiveness against AIDS. To fix our thoughts, suppose it is the case that half the drugs have no effect, and half are effective. (Of course, one would not know this in advance, but it provides a useful reference for the following "thought" experiment.) In this long series of drug tests, there will periodically be cases where the observed significance level (or P-value) is approximately 0.05. To be specific, consider those cases where the P-value, corresponding to the null hypothesis of no effect, has a value between 0.04 and 0.05. What fraction of the corresponding drugs are ineffective?

To recapitulate: we began by knowing that half the drugs were ineffective, but not which half. Now we are looking only at those drugs where the P-value turned out to be about 0.05; what fraction of these will be ineffective? The answer is – at least 25% and typically over 50% ! (The exact answer depends on the degree of effectiveness of the effective drugs, but for typical scenarios the exact answer will be over 50and Jefferys and Berger, 1992.)

It is clear that there is rampant confusion on this point, as Professor Freedman points out in footnote #18. My claim is that the fault lies with the statistitians, for promulgating a concept that has almost no hope of being properly interpreted. Had we all along been advancing the far simpler Bayesian notions (here, a direct statement for each drug test that "the probability that the drug in ineffective is XXX") we would be much further along in statistical understanding. (My argument has not addressed the difficulties in carrying out the Bayesian analysis, or in arriving at reasonable inputs for the prior probabilities, but Bayesians feel that these are easily surmountable hurdles.)

Frequentist versus Bayesian statistics. I have relabelled the debate which Professor Freedman has labelled as the "objectivist – subjectivist" debate for a number of reasons. The primary reason is that there is a very strong objectivist Bayesian school, founded by Simon Laplace in the late eighteenth century, and more fully developed by Harold Jeffreys in the first half of this century (see Jeffreys, 1939). This school tends to interpret probabilities subjectively (although members of this school typically are willing to accept any kind of probability), but uses, as inputs to the analysis, "objective" prior probability distributions. Attempting to be more precise here would serve little purpose; suffice it to say that this school can be argued to be every bit as objective as is the frequentist (classical) school. (Indeed, a large number of our standard classical procedures were first derived in the nineteenth century using the objective Bayesian method of Laplace.)

The debate between members of this school and the frequentist school tend not to revolve around axiomatic issues, but rather around operational and interpretational issues. (An example of the latter is the earlier-discussed issue of interpreting significance levels.) On the operational side, Bayesians argue that they can be much more flexible in their modelling and in the types of questions they can answer. This is , of course, a hotly contested argument, but its existence should be noted.

As an aside, it is of interest to note that many pure subjectivist Bayesians argue almost as vehemently against the objectivist Bayesians as against the frequentists. This is because objectivist Bayesian analysis is not fully compatible with the rationality arguments that motivate the subjectivist Bayesians. (But, as Professor Freedman mentions, if one enlarges the rationality scenario to accommodate practical limitations of time and resources, deviations from the strict position may be justifiable.)

In practice, these distinctions start to blur. In a recent applied analysis there was a part of the problem that was rather data-rich, and for which I used an objective Bayesian analysis. (It was crucial to do a good job of modelling, but the prior probabilities made little difference in the answers, and hence we made no attempts to ascertain what they were.) Another part of the problem had almost no data, however, and for this part an extensive subjective Bayesian analysis (involving elicitation of prior opinions of dozens of engineers) was undertaken. (I will even admit to using a bit of frequentist analysis in a third part of the problem.) The advent of powerful computing capabilities has led to a great upsurge in practical application of the Bayesian method, and practical Bayesians tend not to be too concerned about the subjectivist-objectivist labels.

Conclusions Professor Freedman has focused on perhaps the two most serious foundational issues in statistics, the issue of routine utilization of statistical modelling, and the frequentist-Bayesian issue. It is interesting that our experiences have led us to reach somewhat different conclusions about modelling. While mistrustful of other people's models (and even occasionally, mistrustful of my own), I feel that statistical modelling cures more problems that it creates. Of course, even if Professor Freedman feels the opposite, we can agree on his primary message of the need to increase the wariness towards models. On the Bayesian issue, I agree with Professor Freedman as to the limitations of the rationality arguments and the resulting conclusion that one cannot "prove" subjectivist Bayesian analysis to be the best practical approach to statistics. On the other hand, I feel that there are strong practical arguments for adopting a flexible mixture of the objectivist and subjectivist Bayesian schools, as the basis for statistical analysis.

References

Berger, J. and Sellke, T. (1987), Testing a Point Null Hypothesis: The Irreconcilability of P-values and Evidence (with Discussion). *Journal of American Statistical Association* **82**, 112-139.

Jefferys, W. and Berger, J. (1992), Ockham's Razor and Bayesian Analysis. *The American Scientist* **80**, 64-72.

Jeffreys, H. (1939), Theory of Probability (third edition, 1983). Oxford, Clarendon Press.

E. L. Lehmann
Department of Statistics
University of California
Berkeley, USA

FOUNDATIONAL ISSUES IN STATISTICS: THEORY AND PRACTICE

Abstract. The foundational issues discussed by David Freedman are examined from the point of view of an investigator faced with the choices they imply. It turns out that in practice the boundaries between the various philosophies are less sharp than might be expected from the passion with which their differences have been argued for a long time. There is considerable common ground, and each approach gains by borrowing tools from the other.

1. Introduction

David Freedman's paper consists of two parts. The first describes the frequentist and Bayesian approaches to probability and statistics, the second is concerned with the validity of statistical models. Both emphasize the importance of putting realism above convenience and dogma. I shall here consider these issues from the point of view of an investigator faced with the choices implied by Freedman's distinctions, namely between

- data analysis without a model;

- a frequentist model involving unknown parameters;

- A Bayesian model which assigns a known distribution to these parameters. (I am restricting attention here to what Freedman calls the "classical" Bayesian position; bridge building to the "radical" position is more difficult).

2. Model-free data analysis

At this stage, a data set will typically be examined, analyzed, and organized in many different ways in an attempt to bring out its salient features and to pinpoint meaningful effects. However, the question then arises whether these effects are real. Unfortunately, except in very special circumstances (for an example see Freedman and Lane (1983)), this question cannot be answered without a probability model. The difficulty is greatly exacerbated by the multiple facets of such an approach. The suggested effects of interest, for example, will tend to be those that appear most significant out of large numbers that were examined, some in detail, some only out of the corner of an eye, and many of which – in a phrase of Cournot (1843) – "have left no traces". The appearance of their significance may thus be greatly exaggerated. Cournot thought that the problem posed by this distortion was insoluble. However, often the situation can be saved by a second stage with new data at which a probability model is used to test the much smaller and more clearly defined set of effects found interesting at the first stage.

3. The modeling decision

The choice between an analysis with or without model depends largely on the extent of our knowledge concerning the underlying situation. Do we know enough from past experience and/or a theoretical understanding of the situation to be clear about the questions to be asked and to be able to formulate an adequately realistic model? Or do we have to examine the data to see what the problems are and to get an idea of the kind of model that might be appropriate?

In this connection it is important to realize that models used for different purposes need to meet different standards of realism. A well known classification (see for example Box, Hunter and Hunter (1978), Cox (1990) and Lehmann (1990)) distinguishes between on the one hand explanatory or substantive models which try to portray in detail the principal features of the situation and to explain their interrelations, and on the other empirical or off-the-shelf models, which employ the best fitting model from a predetermined standard family. The latter have little explanatory power but nevertheless can be very useful for limited practical purposes such as predicting school enrollments or forecasting elections. (Early astronomers were able to make very accurate predictions based on elaborate models [1] that bore little resemblance to reality. Their insistence on circular motions is reminiscent of the ubiquitous assumption of linearity in the kind of regression model discussed by Freedman in Section 5 of his paper.) As Freedman points out, the success of such predictions could and should be routinely monitored. A danger of empirical models is that once they have entered the literature, their shaky origin and limited purpose have a tendency to be forgotten.

4. Model-based inference

Suppose now that a model has been selected which we believe to be adequate, and that we know exactly which questions we want to ask. Let us assume that the problem has been formulated as a choice among a set of possible decisions and that we can assess the loss $L(\theta, d)$ that would result from taking decision d when θ is the true parameter value. The observations are represented by a random quantity X which has a probability distribution depending on θ. A statistical procedure is

---

[1] A recent account of the accomplishments and ultimate failures of these models can be found in North (1994).

a rule (or function) δ which specifies the decision $d = \delta(x)$ to be taken when X takes the value x. Finally, the performance of the decision rule δ is measured by the average loss resulting from its use,

$$R(\theta, \delta) = EL[\theta, \delta(X)] \tag{1}$$

This expected loss is called the risk function of δ.

A classical Bayesian has a prior distribution W for θ; for the sake of simplicity we suppose that W has a density $w(\theta)$. Within this Bayesian framework, the quantity of interest is the expected risk

$$r_\delta = \int R(\theta, \delta) w(\theta) d\theta \tag{2}$$

and the most desirable procedure is that for which r_δ is as small as possible. Determining this optimum procedure is in principle straightforward, although it may be computationally difficult.

The hardest part often is the choice of W. In practice, prior distributions, just like many probability models, frequently are of the off-the-shelf kind, that is, they are chosen from a convenient standard family which is sufficiently flexible to permit modeling the rough qualitative features desired. A popular choice is an uninformative prior that corresponds to a state of ignorance concerning θ.

Frequentists do not face the issue of determining a prior, but they have to contend with another difficulty. Rather than minimizing (2), they seek a procedure that minimizes $R(\theta, \delta)$ for all θ. Since, unfortunately, such a procedure does not exist, various devices are used to work around this problem, none of them really compelling. We mention here only one such criterion: minimizing the maximum risk. A procedure achieving this is said to be minimax.

5. Common ground

The distinctions between the three approaches: (i) model-free data analysis, (ii) frequentist and (iii) Bayesian model-based inference have been heatedly debated for a long time. However, in practice there is more contact – and the lines separating the three modes are less sharp – than this debate suggests.

For example, an exploratory data analysis may perform informal tests with ad hoc models, for guidance on which hypothesis suggested by the data to pursue; conversely, the subsequent formal-inference stage may utilize the first stage to search for the models it needs.

We shall in the remainder of this section restrict attention to some of the relationships between (ii) and (iii). Additional discussion of these and other points of contact between the two approaches can be found in Diaconis and Freedman (1986) and Lehmann (1985).

(a) Previous experience. For a large class of important situations in areas such as medicine, agriculture, business and education, a great body of related earlier experience is available to draw on. This experience provides an observed frequency distribution of θ's of which we have reason to believe that the present (unknown) θ-value is a typical member. A Bayesian is likely to accept this distribution as an appropriate prior (although possibly making slight modifications to allow for special features in which the present situation might differ from the past). A frequentist, even without a clearly defined random mechanism for producing the θ's, may analogously work with the empirical distribution as if it were the true frequency distribution of a random quantity. Although they will think very differently both of this distribution W and of the assumed distribution of X given θ (along the lines indicated in Sections 2 and 3 of Freedman's paper), both will be led to the procedure δ minimizing (2) and thus to the same decision.

(b) A fundamental connection. Even when no prior information is available, Bayes procedures are of importance to frequentists. This is a consequence of a fundamental theorem of Wald, which states that the only sensible (from a frequentist point of view) procedures are those that are Bayes solutions for some prior W or limits of such solutions. In the light of this result, it is often useful to examine a frequentist procedure from a Bayesian point of view to see whether it is Bayes with respect to some prior and if so, whether this prior seems at all plausible. In this connection it is interesting to note that minimax procedures are often Bayes solutions corresponding to an uninformative prior.

(c) Performance. In the opposite direction from (b), a Bayesian, after having computed the Bayes procedure δ corresponding to a preferred prior, can learn much about the procedure by studying its risk function $R(\theta, \delta)$. This latter step, although inconsistent with a Bayesian philosophy, is now acknowledged as useful by many Bayesians. (For a discussion from a contrary viewpoint, see Berger (1986).)

5. Conclusions

1. It seems clear that model-free data analysis, frequentist and Bayesian model-based inference and decision making each has its place. The question appears not to be – as it so often is phrased – which is the correct approach but in what circumstances each is most appropriate.

2. In practice, the three approaches can often fruitfully interact, with each benefiting from consideration of the other points of view.

3. Although the debates between adherents of Bayesian and frequentist philosophies have been carried on with much heat for a long time, in practice neither side very often lives up to its principles, since both models and priors frequently are of the off-the-shelf variety.

References

Berger, J. ,1986, Discussion of Diaconis and Freedman, *Annals of Statistics* **14**, 30-37.

Box, G. E. P., Hunter, W. G., and Hunter, J. S., 1978, *Statistics for Experiments*. New York, Wiley.

Cournot, A. A., 1843, *Exposition de la Théorie des Chances et des Probabilités*. Paris, Hachette

Cox, D. R., 1990, Models in Statistical Analysis, *Statistical Science* **5**, 169-174.

Diaconis, P. and Freedman, D., 1986, On the Consistency of Bayes Estimates, *Annals of Statistics* **14**, 1-67, (with discussion).

Freedman, D. A. and Lane, D., 1983, Significance Testing in a Nonstochastic Setting, pp.185-208 in P. J. Bickel, K. A. Doksum, and J. L. Hodges, Jr. (eds), *A Festschrift for Erich L. Lehmann in Honor of his Sixty-Fifth Birthday*. Belmont, Calif., Wadsworth.

Lehmann, E. L., 1985, The Neyman-Pearson Theory After Fifty Years, pp.1-14 in L. LeCam and R. Olshen (eds), *Proceedings of the Berkeley Conference in Honor of Jerzy Neyman and Jack Kiefer*. Vol. 1. Monterey, Calif., Wadsworth.

Lehmann, E. L., 1990, Model Specification, *Statistical Science* **5**, 160-168.

North, J., 1994, *Astronomy and Cosmology*. London, Fontana.

Paul W. Holland
Graduate School of Education
University of California
Berkeley, CA 94720 - 1670, USA

SOME REFLECTIONS ON FREEDMAN'S CRITIQUES

Key Words: Frequentist, Subjectivist, Error-term models, Data analysis, Regression models, Statistical practice

Abstract. Freedman's version of objectivism is criticised along the usual subjectivist lines coupled with the more positive reactions of the applied statistical eclectic. In addition, it is argued that the 'error term' interpretation of regression models that Freedman either takes as a given or regards as an innocuous *faceon de parler* is, in fact, the pebble that inevitably leads to the veritable avalanche of misuses of regression models for causal inference in economics, political science and elsewhere, that he deplores.

I find myself in agreement with much of what Freedman says. I think that we both regard statistical practice as the acid test for the foundations of statistics. 'Foundations' that conflict with seventy-year-old practical lore are to be regarded with suspicion rather than to be embraced simply because of their clarity or logical consistency. I will comment on a few of the many items in his presentation, and include a few diversions here and there that his text suggests to me. Regardless of our differences of opinion, I doubt that Freedman and I would disagree on the analysis and interpretation of a set of data. We might, however, explain our reasoning in different terms.

The Objectivist Position. What is probability? Freedman says that, for an objectivist, probability is an inherent property of the system being studied. Later on he says that for an objectivist like himself the probability of a 'head' in coin tossing has "its own existence, separate from the data" but the data can be used to estimate and to test hypotheses about this probability. None of this really says what probability *is*, which is the usual subjectivist criticism of the position. But ignoring this aspect of the Freedman's presentation, is any system 'really' probabilistic and therefore govern by 'inherent' probabilities? For the world that we perceive I don't think so. Coin tossing' is a physical process whose outcome is (a) completely determined by various laws of motion that are well understood but (b) very sensitive to *initial conditions* that are sufficiently *unknown* in many circumstances that the outcome, though deterministic, looks 'random'. Without being more specific, let me simply assert that many 'random' phenomena are *completely deterministic* in the sense that coin tossing is – e.g., roulette wheels, turbulence, city traffic, random number generators, chaotic non-linear systems, waiting for service in queues, card shuffles, the action of wild fires, the discovery of oil wells, the behaviour of rats in mazes and the responses of humans to test questions. I did not invent this view, of course. 'Ignorance of initial conditions' as a source of probability has been with us since the dawn of the application of the mathematics of probability.

However, saying this does not mean that *treating* these systems as deterministic is a good idea in practice. For many purposes and from many points of view such systems act as though they *are* stochastic and for many purposes a statistical treatment of them is not only useful, it is often the only practical way of proceeding. No matter what other uses it might have, ontology does not entail methodology.

At the quantum mechanical level. I am told by those who say they understand these things, 'real randomness' does exist. I do not claim to understand what this assertion means and I can not judge it, except to observe two things. First, to this day there still are attempts to explain the randomness of quantum mechanics via 'deterministic' process e.g., Bohm and Hiley (1993). Second, whatever the success or failure of such attempts, the fact remains that the 'probability' of quantum mechanics is not like the proportions and relative frequencies of everyday life. It needs negative and imaginary 'probabilities' in order for it to fit the data as well as it is known to do. I do not feel at all abashed about ignoring the 'probability' of quantum mechanical systems in my view that *nothing* we can actually perceive is 'really' a probabilistic process. The RAND (1955) table of random numbers was initially generated by a physical probabilistic process involving quantum mechanical uncertainty, but the results were manifestly not random and various alterations of the tables were needed before they would look 'random enough' to be of any use to those who use such tables. While this story has much to do with the gulf between engineering and basic science it is a lesson that *theories* about randomness based on physics do not necessarily pan out.

As far as I can tell, the 'objectivist' position, either as briefly summarised by Freedman or as given by others, says *nothing* about the *nature* of probability nor has it much to do with being 'objective' nor is it about things that 'objectively exist'. I prefer the other term, 'frequentist', because concern about the relationship of long-run relative frequencies in data to 'probabilities' computed from mathematical models is the *sine qua non* of 'objectivists'. Most frequentists are content to calculate probabilities from certain types of mathematical models and, when they feel that these models are sufficiently applicable to some real situation, they are then willing to make assertions about various empirical large-sample (or "long run") frequencies based on these calculated 'probabilities'.

That said, in my opinion, it is usually quite foolish to ignore a large frequentist probability about an event – whether the event is observable, like the tosses of a coin, or unobservable, like the parameter content of a confidence interval. This admonition goes double if the probability model looks like it is a fair representation of the real phenomenon being studied. How I assess the 'fairness' of such representations is my subjective business, of course.

However, when we are faced with an upcoming singular event, the issue of its probability may *seem* interesting but unless it is a large (or small) probability it may not be all that useful in practice. As we are about to toss a coin what does it mean that its probability of a head is .5? Not much in practice. It will come up heads or tails and we won't be surprised by either occurrence. As we place our bet on 17 at the roulette wheel we know that 17 is probably not going to come up, but if we are very lucky (odds of '37 to 1' worth of 'luck' to be exact) it will! As we contemplate a 95% confidence interval based on data from a well-executed random sample we know there is a chance that the sample data are very misleading, but at 5 chances out of 100 who wants to depend on such things?

I hope it is obvious that then when it comes to the *meaning* of probability I am of the 'degree of belief' camp since I haven't figured out any better interpretation and I feel uncomfortable with giving it *no* interpretation as I believe most frequentists do. As far as the importance of the *meaning* of probability goes, I'm not sure that it is always that important for the practice of statistics.

While I like the radical subjectivists' emphasis on observable or potentially observable data, I do not see how we can ever make progress in practice without invoking unobservables in various ways. The problem is, in my opinion, that, like the emperor, it is all too easy to get very wrapped up in unobservables and to forget that unobserved quantities without empirical consequences are scientifically worthless, no matter what the story is that goes with them. If I have to choose between Freedman's two subjectivist camps I guess I go with the classical group, but as stated (and too often expressed in print) that position is too self-congratulatory for my taste. Statisticians are so vilified by others ("I'm no statistician, but …") that it has always amazed me that they will vilify each other.

The tidy idea of 'coherence' has never appealed to me after I realised that it often means that you can't learn anything from data analysis that you didn't already explicitly assume before you started the research. That is simply ridiculous. If you didn't put some prior probability on a curvilinear model you will never discover the need for one by using Bayes' theorem; but a plot of the data will show it to you in a flash.

On the other hand, I think that problems with many similar parameters often benefit from regarding these parameters as 'random' so that long-run relative frequencies can be brought into play to estimate the values of these parameters.

'Random parameter' models were once the property of the subjectivists but little used in practice until the frequentist results of James and Stein (1961). These results can be read as showing that Bayesian estimates in certain large multi-parameters problems often have a large frequency probability of really 'doing better' than simpler coordinate-wise methods. (This is just one of the reasons that I do not regard it as wise to ignore a large frequentist probability).

One of my favourite examples of something that seems to be intermediate between data and a parameter is the probability content of the empty cells in a contingency table. Robbins (1968) gives a neat frequentist solution to 'estimating' this quantity, but I think this is also good place to think about the two subjectivist camps. We don't know which cells are empty until we have the data in hand. Yet, we often know in advance (from a large frequentist probability) that we will get some empty cells due to the size of the sample. The radical subjectivist might reformulate the problem as finding the probability that we will next observe a value as yet unobserved. Others of us will want a posterior interval on how much of the population distribution is in the sample's empty cells. This is a problem where the distinction between discrete and continuous distributions is large. No matter how much data you have, for a continuous distribution *all of it* (almost surely) *is unobserved*! In my view, this is just another example where it is best to start with discrete distributions in order to organise your thoughts because continuous distributions are usually best regarded as only approximations to the real discrete distributions of life and they can confuse rather then help one's thinking about the issues faced in data analysis, Holland (1979). Once this exercise has become natural, the real, simplifying role of continuous distributions comes more sharply into focus, as it was originally intended.

Probability versus relative frequency. I love the $(TH)^{10}$ example that Freedman uses because it raises so many issues. If the data had only involved four tosses, $(TH)^2$, then I don't think we would be too interested in the alternating pattern. With 20 tosses our intuition(and a small frequentist probability) says that simple coin tossing is *implausible* here. This is a good example of why incoherence is a routine state of grace for a data analyst. A subjectivist who had set up a simple binominal model with a prior for p would not see the alternating pattern since it is irrelevant from the point of view of the posterior of p and thus it is irrelevant for his prediction of the outcome of the 21st toss. However, looking at the data is what gets this example going. The data summary of the number of heads out of 20 tosses is not enough to warn us that there is anything of interest here. What would we do if the example were $(TH)^9HH$?

However, though Freedman offers this as an example of when relative frequency and probability are not the same thing, he continues to fail to tell us what probability *is*. Since I accept the degree-of-belief interpretation I do know what it is and I would agree with him that probability and these relative frequencies are not the same.

'Natural' priors are not the only things that generate their own technical litera-ture. On the frequentist side I would point to the vast churning out of papers on the many aspects of confidence intervals and especially of 'simultaneous inference'. Ex-cept when they can, with impunity, be confused with subjectivist posterior intervals I am forever suspicious of confidence intervals. A fairly recent example of such a suspicious confidence interval is the one proposed by astronomers (it made the New York Times!) for the 'remaining life of the universe'. They assume the probability model that X, the *current* age of the universe, is uniformly distributed on $(0, T)$, where T is the unknown *ultimate* age of the universe. They 'justify' this probability model for X by the assertion that there is nothing special about either these times or our universe. Consequently, $T - X$, the 'remaining life of the universe' is, with 95% confidence, bracketed between $X/39$ and $39X$. This is one of the few large frequentist probabilities that I *will* ignore.

Statistical models. Here I am the most concerned about Freedman's presenta-tion since it is the standard way that people talk about regression models and he perpetuates what is to me the most embarrassing notational confusion for which the field of statistics is responsible – 'error term' models. He is not responsible for this confusion of course, but since he is concerned about the use of these models in the social sciences I think he ought to acknowledge that by appearing to buy into this notation and the story that goes with it (i.e., his "box of tickets", etc.) he only reinforces the misuse of regression models by the very social scientists he is criticising. They think he is one of them!

I am speaking, of course, about the equation: $y = a + bx + \varepsilon$. What does it mean? The only meaning I have ever determined for such an equation is that it is a shorthand way of describing the conditional distribution of y given x. It says that the conditional expectation of y given $x, E(y \mid x)$, is $a + bx$ and the conditional variance of y given $x, Var(y \mid x)$, is the variance of the distribution of ε. For subjectivists it is very natural to talk about conditional distributions but frequentists often seem to have a perverse desire to avoid such references, if at all possible. For example, on the one hand, Freedman uses the conditional expectation interpretation explicitly in his interpretation of his equation (4), but on the other hand, he invents a 'double bar' notation to mean "computed on the assumption that" to let frequentists avoid saying "conditional on" in his discussion of tests of significance.

A standard interpretation of $y = a + bx + \varepsilon$ is that to obtain this value of y we start with a given value of x, compute $a + bx$ and then add the random quantity ε to the result to get y. This is curiously viewed as how x 'causes' y. As Freedman agrees, no statistician has ever argued that the values y_i are *actually* generated this way (except in computer simulations). It is just that they may be conveniently considered to be so generated. In order to do this, the two observables, x and y, must be augmented by the unobservable 'error term' ε. If the joint distribution of the triple (y, x, ε) were exactly equivalent to describing the conditional distribution of y given x there would be no problem, but this is not so. Untestable assumptions must now be made about ε and about its relation to x in order for the equation $y = bx + \varepsilon$ to imply that $E(y \mid x) = a + bx$, $Var(y \mid x) = \sigma$, etc. The necessary assumptions are inherently untestable because ε is unobservable. Furthermore, error-term models like this one and its many generalisations have spawned a huge literature in econometrics and many other social sciences, which I think Freedman deplores (with me) down to the last paper. Once ε is there to pick on it can be given all sorts of untestable properties to keep the word processors busy. What I dislike about these models is that they combine features of the problem that can be dealt with more or less easily, e.g., the form of a conditional expectation or conditional variance as a function of x, with much more subtle 'untestable' features, such as the effect of randomisation or of selection bias. All these items become rolled into 'model specification'. I like to separate these things since some are best settled by data analysis and others are, in the end, purely subjective, i.e., 'settled' by *making untestable assumptions* (which we all do every day and of which we should not be ashamed). Let me now address the Hooke's Law and the salary discrimination examples, in turn.

Hooke's law. If spring s is suspended and weight x is attached to it then it will stretch to length $Y_s(x)$. If $Y_s(x)$ is assumed to be a smooth function of x then, for a range of small values of x, $Y_s(x)$ will be linear, i.e., $Y_s(x) = a_s + b_s x$. The values of a_s and b_s can vary from spring to spring and the equation will hold for spring s and for values of x that are not too large (or too small). The value, $Y_s(x)$, is a hypothetical and unobservable quantity that lies behind actual *measurements* of the length of the spring. If we were to take a real spring, apply a series, x_i, of known weights to it and measure the resulting length, y_i, we would get a series of pairs of observed values, (y_i, x_i). If x_i were set at x and repeated measurements of the length, y_i, were made they would tend to average around $Y_s(x)$. I think this is all that is meant by the equation $y_i = a_s + b_s x_i + \varepsilon_i$. Again it has a simple conditional distribution interpretation.

The quantity $Y_s(x)$ is *causal* in the sense that it tells us what will happen to the length of the spring for all weights in a given range, no matter *when* or in what order we apply them to the spring. This is an approximation of course. Each time a weight is applied the spring is distorted to some degree and does not return exactly to its original length – again, as Heraclitus observed, we can't step into the same river twice. But for a range of weights this effect might be small enough to be ignored. Such 'homogeneity' assumptions, Holland (1986), while common in the physical sciences, are very difficult to believe in many social science applications. They are, of course, never *directly* testable in any setting.

Salary Discrimination. Elsewhere, Holland (1988), I have said my say on the problem of causal inference in salary discrimination cases. It is a problem that is rife with controversy. Freedman uses it as a convenient "stylised example". I believe that in both the Hooke's Law example and in this example the estimation of linear models, as forms for the conditional expectations, is not very controversial and similar methods would be useful in both examples. I mean this to include assessing 'statistical significance' and the like. Freedman seems to disagree because he says that assessing 'statistical significance' is difficult and involves untestable assumptions – "Discrimination cannot be proved by regression modelling unless statistical significance can be established, and statistical significance cannot be established unless conventional presuppositions are made about unobservable error terms".

What I think *is* different about the two examples is the ease with which we can make, and find data to support, assumptions about *counterfactuals*, an important class of untestable assumptions. As usual, counterfactuals play a key role in causal inference. For Hooke's Law one counterfactual of interest is the length of the spring that we would have measured if we hadn't already subjected it to some other previous weights in the process of running the experiment. We are probably willing to assume that the order of applying the weights doesn't matter even though it does matter to some small degree. In the salary discrimination example it is easy to get confused as to what the counterfactual of interest really is. For example, it is *not* what a woman's salary would have been had she been a man with the same qualifications, even though it is the coefficient on the gender variable that attracts all the interest (and we even talk about 'holding qualifications constant')! Rather we want to know what the salaries of men and woman would be if there were no discrimination. This is the counterfactual that Freedman alludes to and almost describes when he says "if the company discriminates, that part of the model can not be validated at all". He is not confused about the relevant counterfactuals, but others can be and discussions about whether 'gender can be a cause' can go on and on, Glymour (1986).

In the Hooke's Law example we can obtain various types of data (involving *s* and other springs as well) to help us examine the 'order doesn't matter' assumption. In the salary discrimination example the data are all collected in an allegedly discriminatory system so we have *no* data relevant to what the salaries would have been under a system of non-discriminatory salary administration. It is not simply a matter of establishing the statistical significance of the coefficient on the gender term in the regression equation – that is non-controversial but might involve complicated computations. It is the *interpretation* that this coefficient is determined by discriminatory practices that is fraught with untested and mostly untestable assumption about counterfactuals. In Holland (1988), I tried to make these assumptions explicit in a simple case and did not find them to be very simple at all and not easily addressed by data.

By continuing to focus attention on untestable assumptions about the unobservable 'errors' in error-term models Freedman gives credence to the view that these models really are the right way to look at things. By emphasising the importance of the assumptions needed to assess statistical significance in such models, rather than on the importance of exposing what the relevant counterfactuals are and how we might try to assess things about them, Freedman gives the *possible* illusion that all we really need is another theorem in probability theory and the problems will be solved. To commander his last sentence and use it for my own ends: "This position seems indefensible, nor are the consequences trivial. Perhaps it is time to reconsider".

References

Bohm, D., and Hiley, J. (1993), *The Undivided Universe. An Ontological Interpretation of Quantum Mechanics.* New York, Routledge.

Glymour, C., (1986), Comment: Statistics and Metaphysics, *Journal of the American Statistical Association* **81**. 964 – 966.

Holland, P.W., (1979), The Tyranny of Continuous Models in a World of Discrete Data, *IHS Journal*, **3**, 29 – 42.

Holland, P. W., (1986), Statistics and Causal Inference, *Journal of the American Statistical Association* **81**, 945 – 960.

Holland, P.W., (1988), Causal Mechanism or Causal Effect: Which is Best for Statistical Science, *Statistical Science*, **3**, 186 – 188.

James, W., and Stein, C. (1961), Estimation with Quadratic Loss, *Proceedings of the 4th Berkeley Symposium* **1**, 361 – 379.

The RAND Corporation (1955), *A Million Random Digits and 100,000 Normal Deviates.* New York, The Free Press.

Robbins, H. (1968), Estimating the Total Probability of the Unobserved Outcomes of an Experiment, *Annals of Mathematical Statistics* **39**, 256 – 257.

Clifford C. Clogg
Departments of Sociology and Statistics
Pennsylvania State University
PA 16802 - 6211, USA

COMMENTS ON FREEDMAN'S PAPER

It is a pleasure to comment on Professor Freedman's provocative paper on the foundations of statistical reasoning.

One of the main tasks of statistics is to summarize uncertainty in inferences. In a given analysis it is important to recognize what inferences are desired (or possible) and also what types of uncertainty are inherent in the analysis. In the social sciences, there are many sources of uncertainty in addition to "sampling error." Incomplete or missing observations, attrition in panel studies, model selection, complex sampling schemes, and biased sampling all create uncertainty (Clogg and Dajani, 1991; Clogg and Arminger, 1993). Statistical methods ought to be tailored to the types of uncertainty most relevant in a given analysis.

It is common to distinguish, as Freedman does, between so-called objectivist (i.e., frequentist) and subjectivist (i.e., Bayesian) approaches to measuring uncertainty. Although the axiomatic foundations of these two approaches are very different from each other, it is often the case that these differences do not have *important* consequences for the summary description of uncertainty. This can be illustrated with an example that is similar to examples given in Freedman's paper.

Suppose that Y is the number of "successes" and $n - Y$ is the number of "failures" in n independent trials. The data consist of observations like coin tosses. The frequentist proceeds by supposing that there is a "true" probability of success for each trial, say π, so that Y follows the binomial distribution; $\hat{\pi} = Y/n$ is the natural estimator of π. The uncertainty in this estimator would typically be summarized as an interval, say $\hat{\pi} \pm 1.96(\hat{\pi}(1 - \hat{\pi})/n)^{1/2}$. It is impossible for the frequentist to interpret the interval obtained in a given sample (say, (.30, .50)), but it is possible to give a precise probabilistic interpretation to the set of interval estimates that would be obtained "in repeated samples." (In a large number of repetitions, approximately 95% of the intervals so constructed would contain the true π.)

Using the same data, a Bayesian would usually consider a noninformative (beta) prior for the unknown parameter and then calculate summaries of the posterior distribution given the prior and the binomial likelihood used implicitly in the frequentist approach. This prior amounts to adding 1/2 to the observed frequencies

(Clogg *et al.* 1991). In effect, the original data $(Y, n - Y)$ are replaced by augmented "data" $(Y + .5, n - Y + .5)$. If it is assumed in addition that the inference is to take place on the logit scale (Bayesian arguments are not required for this), the mode of the posterior distribution is $\log[(Y + .5)/(n - Y + .5)]$, which implies $\pi^* = (Y + .5)/(n + 1)$ as the estimator of π. The (approximate) variance of the logit is $[(n + 1)\pi^*(1 - \pi^*)]^{-1}$, and an (approximate) interval estimator follows from this. The interval estimator for the logit can be converted back to an interval estimator for π. *Frequentist* inference is decidedly better once the data have been augmented by the "prior" frequencies. Samples with zero successes (or with zero failures) are not excluded from the Bayesian analysis, but such samples create special problems for standard frequentist analyses, at least for interval estimation or prediction.

For a given sample, the Bayesian and the frequentist are very likely to come up with the similar summary descriptions of the uncertainty in the inference. Suppose we observe $y = 40$ successes in $n = 100$ trials. The ordinary interval estimate for π is $(.30, .50)$ and the interval obtained by using the noninformative prior gives $(.31, .50)$. The point is that even though the axiomatic foundations that lead to these two intervals differ radically, there is no practical difference between the *summary* descriptions of uncertainty.

If past experience with similar data leads the analyst to suspect that the "true" probability of a success is different from one half, the Bayesian would pick a different prior. If some really informative prior were used (say, add 50 to the number of successes, .1 to the number of failures), inference statements could be affected, at least for samples of small or moderate size. But we would not obtain radically different inferences if the prior frequencies added were, say, $(.9, .1)$. For the example above, this prior leads also to the interval $(.31, .50)$. Frequentists and Bayesians interpret the intervals in very different ways, and they might perform different calculations to summarize either the likelihood or the posterior. In some sense, as Freedman says, the Bayesian who uses a truly informative prior constructs an inference that is logical in his or her own mind only. The standard solution to this dilemma is to either examine the sensitivity of the inferences to the prior, or the sensitivity of the inferences to the selection of Bayesians who analyzed the problem. So even with variability in priors (or variability among Bayesians), an objective inference is possible in principle.

Freedman considers two very different statistical models and asks how different foundational principles might be used to "validate" these models. With the model describing Hooke's law, which I take to be a physical law (or a theoretically known function), it is obvious in principle how inference should proceed. Design an experiment so that the relevant constants can be estimated; collect observations (measurements of lengths and weights) at random; estimate the parameters by least squares or a similar method; assess "sampling variability" (experimental error), here the only relevant source of uncertainty, in terms of (hypothetical) repetitions of the experiment. Valid and objective answers can be obtained using this frequentist approach.

Ordinary Bayesian analysis of this model and similar experimental data, using just about any *sensible* noninformative prior for the parameters, would arrive at virtually the same inference about the physical constants in Hooke's law. And the ordinary Bayesian methods for this problem give "objective" answers, just like the frequentist answers, by which I mean that the answers do not exist only in the head of the experimenter. These answers can be checked objectively by repetition of the experiment.

The situation is different in the application of statistical models in social research. One of the important differences is that experimentation is less common or is less feasible; models are applied most often to nonexperimental data, such as survey data or data on salaries of employees in a firm.

Consider the following three regression models for salary (Y), male $(X,$ coded $X = 1$ if the employee is a male, 0 if a female), and other predictors (say Z):

$$M_0 : Y_i = \alpha_0 + \varepsilon_{i0}$$

$$M_1 : Y_i = \alpha_1 + \beta_1 X_i + \varepsilon_{i1}$$

$$M_2 : Y_i = \alpha_2 + \beta_2 X_i + \gamma Z_i + \varepsilon_{i2}$$

Model M_2 stands for a possibly large set of models depending on which Z variables are included. As in Freedman's paper, suppose that the "sample" consists of all employees in a given firm.

There is nothing inherently Bayesian or frequentist in the definition of these models. But these models can be viewed in several different ways and what it means to validate one or more of these models depends on the view taken. These models can be viewed as descriptive models, as predictive models, or as "explanatory" (or "causal") models. The statistical reasoning that ought to be used depends on the view that is taken, which is an important point.

For model M_0, α_0 is just the mean salary in the firm, and the least squares estimator of α_0 is just the observed mean (in the "sample"). Whether α_0 and its "sample" value ought to be distinguished is an open question. But this model merely reproduces the mean salary and the error term in this model merely represents the fact that not all employees have the same salary. The model is a valid way to represent these facts. Assessing the uncertainty in the sample mean salary is another matter. The observations are surely not independent, and unless one knows how the observations are correlated (as with some measure of intraclass correlation), it may not be possible to give a valid summary description of the uncertainty (say an interval estimator) or even for that matter to define what is uncertain about the sample (population?) mean.

For model M_1, β_1 is just the male–female difference in mean salary, and the least squares estimator of this quantity is just the observed difference in the means. Again, there is little to "validate"; this model is a (valid) description of the gender differential in average salary in the given firm. For this model it is also unlikely that the error terms will be independent of each other, so it is also difficult to assess the uncertainty in the "estimate" of β_1. Finally, for model M_2, the male–female difference in mean salary adjusted for Z (say educational attainment or experience) is β_2. As with the other models, there is little to validate unless we make additional demands of the model. To determine if the coefficient of Male (X) is "significant" takes quite a lot of extra work and additional (perhaps untestable) assumptions, which is one of Freedman's points. A standard modeling approach (whether Bayesian or frequentist) is to assume that if the right Z variables are added, then it can be assumed that the error terms are independent. But even if we could know which Z variables produce this condition, then this only justifies "significance tests" for the regression coefficients in whatever model produces this condition.

The above models are each valid as descriptions of salary "structure" in the firm. The parameters in these models have objective meaning, but it is difficult to defend standard summaries of uncertainty that would be used in the analysis of these models for the given data. Neither frequentist nor Bayesian approaches can make errors independent (or exchangeable) when they are not.

We can ask how well the above models or related models *predict* salary given information on employees. Prediction is inherently objective because predictions can be checked, regardless of how predictions are generated. Perhaps several models of the M_2 type (with several Z variables, not just one Z) would predict well, as can be judged best perhaps by out–of–sample predictions or with fitting the models to data from other firms. Whether priors are helpful or not can be judged objectively by trying out the predictions produced with those priors. But if the observations (or error terms) are not independent (or "exchangeable"), priors do not solve matters; but the point is that the models can be validated as predictive models, and this task would be simplified if error terms could be assumed to be independent of each other.

Suppose that the data collection method were as follows. Instead of sampling all employees in a given firm, randomly sample one member from each of n randomly selected "similar" firms. Suppose further that n is small relative to the total number of "similar" firms. The inference questions change as a result of the data collection. Now the inference pertains to the "population" of firms, for example, and the firm sizes (number of employees per firm) could be taken into account in a variety of ways. With this data collection method, the error terms for *any* model would be independent of each other, and because of this either a frequentist or a Bayesian analysis of *any* regression model would be valid, as judged by the principles of either framework. That is, inferences about the descriptive aspects of those models (the "population" parameters) or about the predictive performance of those models would be greatly simplified, given of course that the model descriptions or the model predictions pertain to a "population" of firms. Either a frequentist or a Bayesian approach could be used to obtain valid (and probably similar) inferences.

Model validation in the sense used by Freedman pertains to the use of the models to estimate causal effects. In language now common in the social and behavioral sciences, the models are regarded as "causal models," not as descriptive models or as predictive models. The logic of "model validation" that Freedman rightly criticizes can be described as follows: (1) pick a model that predicts well, (2) call the regression coefficients in this model causal effects (Clogg and Haritou, 1994). This logic characterizes much empirical social research. In Freedman's example, with the alternative data collection method described above, this logic would say that the (estimated) coefficient for X (Male) is the (estimated) causal effect of X. The significance of this coefficient can be determined with standard frequentist methods. It would be said, in the court room or in social science journals, that employers in such firms discriminate if this coefficient is significantly different from zero for the model that is picked. The basic dilemma is that we cannot know whether the fitted model simulates a controlled experiment or not. Adding priors for the parameters does not solve the problem. Not adding priors does not solve the problem either. The *causal inference* problem is difficult, in spite of the fact that inference about the descriptive or the predictive aspects of these models has been resolved (at least to my satisfaction) by using a different method of data collection.

With data obtained as described above, most would agree that the assumptions justifying either a Bayesian or a frequentist analysis can be maintained for ordinary inferences or summaries of uncertainty in those inferences. That is, either approach has merit as a method for studying the models as descriptions or as predictions with data sets of this general type. But knowing that error terms are independent of each other says little if anything about validating the model as a behavioral explanation or a causal model. We do not know which model of the many that might be picked best simulates a controlled experiment. The primary source of uncertainty for the causal inference (inference about discrimination) is this type of model uncertainty. And this type of model uncertainty is very different from picking the "best" model for the purpose of prediction. A Bayesian approach is helpful, however. The model uncertainty might be taken into account by summarizing the variability in the between–model estimates of the coefficient for X (Male) across all (or many) possible models that might be used (Clogg and Haritou, 1994). It is uncertainty in picking the model, not "sampling error" and not sensitivity of coefficient estimates to priors used, that ought to be investigated. Bayesian perspectives at least allows consideration of sources of uncertainty besides sampling error in these situations, which is one argument in their favor.

References

Clogg, C. C., Rubin, D. B., Schenker, N., Schultz, B., Weidman, L. (1991), Multiple Imputation of Industry and Occupation Codes in Census Public–Use Samples Using Bayesian Logistic Regression, *Journal of the American Statistical Association* **86**, 68-78.

Clogg, C. C. and Dajani, A. (1991), Sources of Uncertainty in Modeling Social Statistics, *Journal of Official Statistics* **7**, 7-24.

Clogg, C. C. and Arminger, G. (1993), On Strategy for Methodological Analysis, pp. 57-74 in P. V. Marsden, ed., *Sociological Methodology 1993*, Oxford: Basil Blackwell.

Clogg, C. C. and Haritou, A. (1994), The Regression Method of Causal Inference and a Dilemma with this Method, forthcoming in S. Turner and V. McKim, eds., *Causality in Crisis?*, University of Notre Dame Press.

Neil W. Henry
Departments of Mathematical Sciences and Sociology
Virginia Commonwealth University,
Richmond VA 23284-2014, USA

THOUGHTS ON THE CONCEPT AND APPLICATION OF STATISTICAL MODELS

Key Words: Statistics, Probability, Model, Regression, Validity, Data analysis.

Abstract. Freedman's arguments about the misuse of statistical models are sound, but I would go further. The mathematical statistician's models are too often taken to be models for substantive theory rather than justifications for examining data in specific ways. Foundational problems relating to the validity of models of theory are the responsibility of the scientists who adopt them and cannot be resolved by statisticians independently of subject matter considerations.

Freedman points out that there are two great foundational problems for the discipline of statistics, namely the way mathematical probability is interpreted and the way statistical models are validated. I propose that one way to eliminate these problems is to give up the notion that statistics is an independent discipline. The problems themselves will remain, of course, but they will no longer underly something called the logic of statistical inquiry. Instead they will be part of the foundations of the disciplines in which probability and statistical models are used. Solutions . . . if that is the right word . . . will need to be phrased in terms of the language and concepts of the disciplines, and they may vary from one to another.

I am going to draw on two sources in writing this commentary. The first is the recent collaborative effort by the historians of science Gerd Gigerenzer, Zeno Swijtink, Ted Porter, Lorraine Daston, John Beatty and Lorenz Kruger, "The Empire of Chance" (Gigerenzer et al, 1989). I'll refer to this as EC for short. The second is a collection of papers presented at a 1960 colloquium on "The Concept and the Role of the Model in Mathematics and Natural and Social Sciences" (Freudenthal, 1961), abbreviated CRM.

Freedman, in this and other papers, contrasts the use of models (specifically regression models) by modern social scientists with the way they were, or might have been, used by physicists working in the mechanical tradition of the 18th century. The fundamental use of probabilities in this tradition arises in descriptions of measurement error. The physical laws are profoundly deterministic; observations are variable. Freedman's use of the spring-and-weight problem is a good example. The statistical model helps us resolve the fact that numerical measurements in the laboratory fail to satisfy the deterministic law, especially as our measurement tools themselves become more refined.

One chapter in EC, "The probabilistic revolution in physics", documents how the role of stochastic elements within physics developed from the 18th to the 20th century. Fundamental to the story told there is the tension between two interpretations of probability. On the one hand probability may be conceived of as something inherent in the objects under study by the discipline, and on the other it may "only serve to characterize our relationship to that theory or to fill preliminary gaps left in the implementation of the ideal program." (EC: 166) The latter view is the one typically expressed in textbooks on applied statistics: the random variable epsilon in the regression model is composed of errors in measurement of the dependent variable and of other factors that have been omitted from the model.

In the physics of thermodynamics and quantum mechanics, however, probability is an essential element in the theory of the phenomena. Furthermore, these areas must wrestle with alternative interpretations of probability: is it defined at the level of the collective (ensemble) or is it a property of each individual element in the collective? "In physics, probability may refer to frequency distributions produced by interactions in large aggregates of similar systems or to a qualified tendency or propensity inherent in elementary constituents composing such aggregates" (EC: 182). (This same ambiguity, of course, exists with regard to sociological theories.) The issue of the appropriate interpretation of the concept, however, becomes one that is fundamental to the foundations of physics, not the foundations of statistics. Resolution, if there is to be any, must be sought within that discipline, using its own language and concepts. If, as it may, a similar problem arises in another discipline, psychology for instance, a different approach to resolution may be necessary.

In another context EC refers to the traditional urns and balls models of mathematical statistics (which Freedman has modernized to boxes and tickets) : "The more the urns and balls are filled with content, the more content- dependent reasoning becomes important" (EC: 228). Patrick Suppes in 1960 had already begun to elaborate this point with respect to psychological theorizing and experimentation (CRM: 173-176).

The concept of model itself turns out to be as controversial as that of probability. Both Leo Apostel (1961) and Suppes (1961) give exhaustive lists of how the term is defined and how models are used in the various disciplines. Their papers remain relevant 35 years later. Apostel focuses on the functions that models serve, concluding that only if we can give a "strict and formal definition for the function of a model" can a unified definition of the concept of model be attained (CRM: 36). His most trenchant observations, however, are contained in a long paragraph on models and explanation. He rejects the idea that "to explain is to infer", based on "the fact that explanation occurs so often, or nearly always, through model-building", the fact that "models are given as explanations of the systems they are models of" (CRM: 14- 15).

Suppes seems to take a stronger position with respect to unified definition: "I would assert that the meaning of the concept of model is the same in mathematics and the empirical sciences. The difference is to be found in these disciplines in the use of the concept." This, even though he points out that "it is very widespread practice in mathematical statistics and in the behavioral sciences to use the word 'model' to mean the set of quantitative assumptions of the theory", and "to confuse or to amalgamate what logicians would call the model and the theory of the model"(CRM: 165). The phrase "models of the data", which has become a commonplace in the ensuing years, is something Suppes finds necessary to explain in great detail. Despite standard textbook statements to the contrary, "the maddeningly diverse and complex experience which constitutes an experiment is not the entity which is directly compared with a model of the theory. Drastic assumptions of all sorts are made in reducing the experimental experience, as I shall term it, to a simple entity ready for comparison with a model of the theory" (CRM: 173).

Where does this get me in regard to Freedman's unhappiness with the use of regression analysis to justify Census readjustment and compensatory salary adjustments for female employees? Quite obviously in these applications of statistical modelling there is little or no substantive theory underlying the models. The abstract model itself has become a theory of data, independent of any behavioral theory. In another piece Freedman (1987) claimed that the widespread adoption of the path analysis or "causal modelling" paradigm had a pernicious effect on a quarter century of social science theory development. I would go even further and say that the institutionalization of statistics as the objective science of data and the icon of scientific method has been much more inhibiting to scientific progress.

Statistics is taught as method, but too often it has become the theoretical language of those who adopt the method. (I include statisticians themselves in this group.) The ubiquity of linear models is a case in point. For every social and behavioral scientist who labors over refining a stochastic model that is faithful to the fundamental theories of her discipline there seem to be a hundred who are satisfied to run their data through a regression or manova machine and claim to be testing theories. Clarifying the distinction between method and model is a challenge that the statistical community itself must address.

I think that Freedman's contrasting examples of regression analysis could be used to do this. First, I need to take the Hooke's Law example out of its obviously experimental context. Instead, imagine that a statistician with no knowledge of mechanics walks into a room whose ceiling is hung with similar looking springs each with a labelled weight rigidly attached. They are stretched to different lengths. The statistician can look, but he better not touch. The simple regression analysis Freedman describes might fit the length/weight data pretty well, though the statistician might point out that better fits could be had by using weight to the 0.94 power or by including a quadratic or cubic term in the equation. In any case, the statistician would walk away convinced that there is a monotonic relationship between weight and length, pleased with his relatively simple "model for the data." Will length change if weights change? Maybe, and maybe not: either call in a physicist with a deeper theoretical understanding of springy matter or look for rooms full of different kinds of data. The regression analysis seems justified as description (the average length of springs with weight X is $a + bX$, more or less) or as a vehicle for prediction.

I'd insist on interpreting the regression analysis of salary data in a similar fashion: a reasonable method of summarizing some facts about a complicated situation. I believe this is what I have done in the two salary analyses I've done: the infamous ASA baseball salary project and a gender equity study for my university. In neither case was there a pretense of developing or testing a behavioral theory of how salaries were set. In the latter case we pointed out to the university administration that if they really wanted to document how that was done they should investigate the people in the university who were responsible for those actions. The role of the statistical analysis was to provide average salary comparisons for individuals with specific sets of characteristics, and the fact that the inclusion of other characteristics in the model might change some of these comparisons was stated up front. The word "significant" was not used. The administration decided to close the statistical "gender gap" through an extraordinary compensation of female faculty. To my mind this was because their "disciplinary" perception of a econo-socio-psycho-logical theory of salary determination, to which the statistician was not privy, made it seem more likely that some salaries had been set because "she doesn't have a family to support" or "she won't take that offer because her husband won't leave his job" than because "she is less qualified for the position."

In conclusion I'll restate my attitude towards Freedman's conclusion. His questions (e.g. "How do statistical models connect with reality?") are vitally important, but they are not answerable by statisticians. Fitting models is informative, and so are standard errors even if they are calculated on an "as if by random sampling" premise. But fitting models as a method of analysis must not be confused with testing theory, or even with (in Suppes' phrase) testing a model of a theory.

References

Apostel, L. (1961), Towards the Formal Study of Models in the Non-Formal Sciences, pp. 1-37, in CRM.

Freedman, D. (1987), As Others See Us: A Case Study in Path Analysis, *Journal of Educational Statistics* **12** 101-223, with discussion.

Freudenthal, H. (1961), [CRM]: *The Concept and the Role of the Model in Mathematics and Natural and Social Sciences.* Dordrecht, Holland, D. Reidel.

Freudenthal, H. (1961), Models in Applied Probability, pp. 78-88, in CRM.

Gigerenzer, G., Swijtink, Z., Porter, T., Daston, L., Beatty, J. and Kruger, L. (1989), [EC]: *The Empire of Chance: How probability changed science and everyday life.* Cambridge, Cambridge University Press.

Suppes, P. (1961), A Comparison of the Meaning and Uses of Models in Mathematics and the Empirical Sciences, pp. 163-177, in CRM.

Foundations of Science
1 (1995/96), 69-83

David Freedman

REJOINDER

Probability theory and mathematical statistics are beautiful subjects, and relatively non-controversial: a theorem is a theorem. Mistakes may be made, even quite subtle ones; but the research process seems in the main to be self-correcting. Applied statistics is a much thornier business because the crucial questions are at the interface between the mathematics and the world: such questions cannot be solved by pure reason. The discussants are to be thanked for the work they have done in trying to clarify the issues. As will be obvious, professional agreement starts breaking down somewhere along the way from mathematical theorizing to practical applications. However, there is consensus on one point: we are all indebted to *Foundations of Science* and its editor, Paul Humphreys, for pulling this discussion together.

In the balance of my rejoinder, I will try to provide some context for the discussion and then address some of the major points made by the discussants. My essay sought to delineate the limits of statistical modeling, particularly regression models in social-science applications. Such models are widely used to make causal inferences – with scant justification. Great confusion has resulted. Modelers often defend their creations on the grounds that the models embody rigorous, state-of-the-art science; in the alternative, that modeling is just standard practice. No discussant made those kinds of arguments. Then there are a variety of tertiary defenses, including the lesser-evil argument: model-based statistical analysis is less bad than analysis without models. Generally, discussants who favored modeling did so on that basis. Thus, I eschew description of more remote lines of defense. In this sort of exchange, proponents of modeling have an easy win, by citing a list of convincing examples. Readers will judge whether any discussants took that path.

Regression equations are used for (i) description, (ii) prediction, and (iii) inferring causation from association. Their efficacy for (i) may be granted, although practical difficulties should be noted. Sometimes, compact description permits substantive conclusions.[1] With (ii), the evidence is mixed. Meehl (1954) argues that models do better than experts in some psychological assessments; I believe he is right, although questions remain about some of the evidence. In economic forecasting, on the other hand, the models at best get a tie.[2] Inferring causation from association is the most intriguing use of the models, and the most problematic; this is where omitted variables, specification error in functional forms, and implausible stochastic assumptions create the most acute difficulties. It is the possibility of causal or "structural" inference that makes regression so popular in the social sciences.

Berger believes that (i) "statistical modeling cures more problems than it creates" and (ii) the Bayesian approach is to be preferred. These opinions are reasonable, but I cannot agree with them.

(i) The superiority of modeling is demonstrated by reference to one example, from Berger's consulting practice. By his account – which I am prepared to accept – Berger's model beat the old fogies. The score is now models 1, fogies 0; but that hardly justifies strong conclusions. Furthermore, I cannot accept the account of the "'good old days' when [medical policy] decisions were made by guesses based on crude data summaries ... anecdotal cases and uncontrolled experiments." Using lots of old-fashioned detective work and well-designed but non-experimental studies, Snow (1855) discovered that cholera was a water-borne infectious disease; Semmelweis (1861) found the cause of puerperal fever; Goldberger (c.1914) found the cause of pellagra; and Müller (1939) identified smoking as the cause of the lung cancer epidemic.[3] The only statistical technology used in these investigations was the comparison of rates. My list could easily be extended. Although a complementary list could also be drawn up, I believe the fogies deserve more credit from Berger and the rest of us.

(ii) Berger prefers the Bayesian position because "the assumptions ... are clear and the conclusions are easily interpretable." This argument is hermetic. Stating

---

[1] For examples, see Freedman (1987, pp.122-23), reviewing work by Blau and Duncan (1967), Hope (1984) and Veblen (1975). Descriptive use of regression avoids standard errors and t-statistics, or interpretation of coefficients as measures of causal effect.

[2] There is more recent work on model-based predictions in psychology, with stronger claims and perhaps weaker evidence, in Dawes (1979). For a brief review of the track record in econometric forecasting and citations to the literature, see Freedman (1987, pp.122-23).

[3] Semmelweis eventually got experimental proof of his thesis. Goldberger's observational studies are classic; some of his diet-intervention studies have proved controversial (Carpenter, 1981); the papers are collected in Terris (1964). There is a very good account of the evidence on smoking, by Doll and Peto, et al. (International Agency for Research on Cancer, 1986).

assumptions is one thing, testing them is quite another. Berger makes no pretense of model validation, he merely argues the lesser evil. Why does he care whether assumptions are explicit? Nor are the conclusions of a Bayesian analysis "easily interpretable," unless you happen to be a Bayesian with the prior used in the analysis.

Berger is an "objectivist Bayesian, [who] tends to interpret probabilities subjectively ... but uses, as inputs to the analysis, 'objective' prior probability distributions." Loosely read, he appears to suggest that subjective probabilities can be derived from the data using only objective, empirical inputs. Of course, that is impossible, as Berger well knows.[4] That is why he has quotes around "objective" in the phrase "'objective' prior probability distributions."

"Objective" priors, like the "uninformative" priors referred to by other discussants, are basically uniform distributions on unknown parameters. If a uniform distribution represents your prior opinion, then a statistician can compute a posterior for you. On the other hand, if that uniform distribution does not represent your prior, the corresponding posterior does not quantify your uncertainties in the light of the data – unless you make it do so by fiat, accepting the inconsistencies that will result. I hasten to add that the prior may have other virtues. For example, Berger and Lehmann (among others) have proved beautiful theorems about uninformative priors. In short, the preference for models and Bayesian techniques seems to be more a matter of personal judgment than objectively demonstrable merit.

Berger makes some welcome points: "we can agree on [the] primary message of the need to increase wariness towards models .... [and] the limitations of the rationality arguments and the resulting conclusion that one cannot 'prove' subjectivist Bayesian analysis to be the best practical approach to statistics." For my part, I agree that Bayesian statistics provides an evocative language, and is a rich source of mathematical questions. In practical applications, if the main elements of a statistical decision problem are in place – the model and the loss function – the Bayesian approach is a powerful heuristic engine for suggesting good statistical procedures. I return to this point and to uninformative priors, below.

---

[4]We wish to compute $P(\text{theory} \mid \text{data})$, where P represents our subjective degree of belief. As good subjectivists, we use Bayes' rule. To avoid technicalities, suppose there are only a finite number of theories and a finite number of possible data sets. Then

$$P(\text{theory}_i \mid \text{dataset}_j) = \frac{P(\text{theory}_i \ \& \ \text{dataset}_j)}{\sum_k P(\text{theory}_k)P(\text{dataset}_j \mid \text{theory}_k)}$$

The denominator involves the marginal distribution of theories with respect to P. So there we go again, with subjective degrees of belief about the theories in advance of data collection. As Jimmie Savage used to say, you can't eat the Bayesian omelette without breaking Bayesian eggs.

Lehmann. Over the years, I have learned a great deal about mathematical statistics from Lehmann, his books, and his students. However, on this occasion, he bypasses some central questions: when is it legitimate to use a statistical model, or a loss function, or a prior?

Lehmann begins by offering a three-way choice: (i) data analysis without a model, (ii) frequentist modeling, and (iii) Bayesian modeling. Missing from this list is a fourth possibility: the data cannot answer the question of interest. This fourth possibility is an important one; statistical models are often deployed to answer the unanswerable questions, by smuggling in crucial assumptions. If need be, those assumptions can be defended later by rhetoric rather than empirical investigation. To be sure, that is not Lehmann's way; but he elects not to discuss the issue.

He reviews the decision-theoretic framework for statistical modeling. In real problems, the loss functions and parametric families of probability distributions are largely unknown – so the statistician makes them up, or chooses them from the stockpile of previous constructions ("off the shelf," as Lehmann puts it). What good does that do?

Lehmann seems to defend formal statistics when multiple looks at the data make it hard to decide whether observed patterns are real, or just reflect chance capitalization. The cure may be worse than the disease. Empirical social scientists routinely fit many models and test many hypothesis before settling on the final equation and its t-statistics. Thus, reported P-values may be largely ceremonial in value – precisely because the search for significance has left no traces, in Cournot's wonderful phrase.[5]

My paper cites misuse of models to adjust the census or compel legislative redistricting; I also cite reviews of models in economics, political science and sociology.[6] Instead of discussing those applications, Lehmann suggests that "in areas such as medicine, agriculture, business and education, a great body of related earlier experience is available to draw on." But he does not give any examples of models built on that experience.

In education, much of the statistical work seems to involve linear models and their variants; for instance, hierarchical models have recently come into fashion. However, there have been serious critiques of linear methods, including Baumrind

---

[5] Dijkstra (1988) discusses chance capitalization.

[6] On census adjustment, see *Survey Methodology*, vol. 18, no. 1, 1992; *Jurimetrics*, vol. 34, no. 1, 1993; *Statistical Science*, vol. 9, no. 4, 1994; these references discuss, among other things, the application of hierarchical models; also see Freedman et al. (1993, 1994). On the use of ecological regression models in voting rights litigation, see *Evaluation Review*, vol. 15, no. 6, 1991; Klein and Freedman (1993). Linear models in political science, sociology, and psychological measurement are discussed in Freedman (1985, 1987, 1991, 1994). Econometric models are discussed in Daggett and Freedman (1985), Freedman, Rothenberg and Sutch (1983). There are well known critiques by econometricians, including Liu (1960) and Lucas (1976).

(1983), de Leeuw (1985), Ling (1983), Luijben, Boomsma and Molenaar (1987), MacCallum (1986), Rogosa (1987, 1993). In medicine, we have learned a great deal in the last century, but do we know enough to justify routine applications of the proportional hazards model? of the statistical models used for carcinogenic risk assessment?[7] Business and agriculture, of course, remain possibilities.[8]

Lehmann's opinions deserve respectful attention. He may at times be more optimistic about modeling than I am: but he does not give strong evidence in support of such optimism. There are, of course, many points of contact. He is right to note positive interactions between objectivist and subjectivist approaches. We can also agree that "empirical" (that is, "off-the-shelf") models "have little explanatory power but nevertheless can be very useful for limited practical purposes .... A danger of empirical models is that once they have entered the literature, their shaky origin and limited purpose have a tendency to be forgotten." Lehmann concludes by warning that "both models and priors frequently are of the off-the-shelf variety."

Holland. My essay outlined the main difficulties with the objectivist framework; Holland is assiduous in supplying detail. He is equally assiduous in dodging the hard questions about his own views. What are subjective degrees of belief, where do they come from, and why can they be quantified? Even the goal of the enterprise is left obscure. Suppose Holland succeeds in quantifying some of his uncertainties. What are objectivists supposed to do with the results, apart from expressing polite interest? Or subjectivists with different priors?

Holland has many concerns about modeling; I respond only to the most urgent among these.

(i) Holland does not quite see what is special about Hooke's law.[9] But the data

---

[7]For reviews, see Freedman and Zeisel (1988), Freedman and Navidi (1989, 1990).

[8]But see Ehrenberg and Bound (1993).

[9]Holland objects to unobservable error terms in regression models; to eliminate such terms, he suggests recasting the models in terms of "conditional expectations." However, this suggestion cannot help. Indeed, expectations apply to random variables. For definiteness, let us take the salary determination model in my paper; let us agree that salary, education, experience (and sex?) are observed values of random variables. We index the employees by i, as before, and write the corresponding random variables as Y_i, X_i, Z_i, W_i. Holland's preferred formulation, which would replace (3) in my paper, is

$$E\{Y_i \mid X_i, Z_i, W_i\} = a + bX_i + cZ_i + dW_i.$$

As will now be clear, he has drawn a distinction without a difference, because the error term ϵ_i is still there in the new formulation:

$$\epsilon_i = Y_i - a - bX_i - cZ_i - dW_i.$$

look like the predictions of the model, and the causal implications can be tested experimentally – as he eventually agrees.

(ii) He is afraid that, in the social sciences, my paper will add "credence to the view that these models really are the right way to look at things." He should relax. Nobody will think that I am adding credence to the modeling enterprise. Subtraction is my game.

(iii) He also thinks that my paper may create the "illusion that all we really need is another theorem in probability theory ...." This too is over-protective. My position should be crystal clear. The connection between models and reality is the central issue. Another theorem cannot help, because we cannot prove by pure mathematics that mathematical models apply to real problems. Empirical reality is not a construct in Zermelo-Frankel.

For Hooke's law and similar physical-science applications, the models seem to be in good shape. With typical social-science applications, regression models are inappropriate and misleading. The difficulties start with the unobservable error terms and grow with the counterfactuals, just as Holland says. His own writings have done much to clarify the latter set of issues. This time around, I focused on the error terms, but I have followed his lead elsewhere.[10]

Holland and I seem to be describing the same elephant; we agree that it is a very sick beast. Our differences, such as they are, come about because I looked at the

---

According to Holland, with his formulation, the assessment of statistical significance "is not very controversial." That is a peculiar claim. As a technical matter, the assessment of statistical significance rides on the assumptions about the ϵ's: for instance, given the explanatory variables, the ϵ_i may be assumed independent and identically distributed with mean 0. Holland does not state his assumptions or defend them – but he expects others to accept the consequences. Why?

As a subjectivist, Holland could take another tack: the joint distribution of the random variables represents his opinion about the problem before data collection; he can "verify" assumptions about the error terms by introspection. So be it. However, significance tests get at questions like the following: is the parameter d in the salary equation zero, or not zero? A subjectivist does not need to test such a hypothesis; he can find the answer by further introspection, because d is now just another feature of the prior. That defense of significance levels will not do.

With Hooke's law, the assumptions can be tested quite rigorously in the laboratory. Among other things, large amounts of data can be generated, so the unobservables become (almost) observable. With the salary example, validation seems much more difficult. Holland's point needs to be reformulated: the treatment of unobservable error terms in the social-science literature has created an enormous amount of confusion. On that, he is surely right.

[10]Holland (1986, 1988, 1994); Allen and Holland (1989); Freedman (1987, 1994). Also see Balke and Pearl (1994), Pearl (1994), Robins (1986, 1987), Robins et al. (1992). For an elementary exposition of statistical models with counterfactuals, see Freedman et al. (1991, pp.462-8). The idea goes back to Neyman (1923), for agricultural experiments.

part of the animal nearest me, while Holland insisted on considering the symptoms most familiar to him (which is by no means unreasonable).[11]

Clogg and I seem to agree quite closely on the modeling issues. "The basic dilemma is that we cannot know whether the ... model simulates a controlled experiment or not. Adding priors for the parameters does not solve the problem .... The causal inference problem is difficult ...," although, as he goes on to say, the models can be useful for description or prediction.[12]

What about the foundations? Clogg argues that subjectivist and objectivist inferences will often give similar measures of uncertainty; the example he gives is coin tossing. He is right, if the inference problem is relatively simple and the amount of data large; that is the content of the Bernstein-von Mises theorem (note 12 to my paper). However, as he remarks too, the interpretations of the numbers remain quite different. An objectivist who has to explain a "95% confidence level" to a subjectivist must squirm, at least a little. (Clogg is quite gentle about this.) Conversely, a subjectivist has problems of his own: why do I care about his 95% probability?

In more complicated problems, subjectivists and objectivists are likely to part company.[13] How much difference would the choice of prior would make in real applications? "Sensitivity analysis" – that is, trying out several priors whose qualitative features conform to the investigators' rough ideas of uncertainty – might give some useful clues, as Clogg suggests. From my perspective, the real problems come in long before such questions are raised, and I believe that Clogg agrees.

---

[11]The elephant metaphor is apt, at least to some degree. When it comes time for practitioners to say just why they are processing the data the way they do, their logic is remarkably opaque – which leaves the rest of us groping for explanations. See, for instance, King, Keohane and Verba (1994).

[12]Clogg suggests that the regression model for salaries would apply if we take a random sample of firms, and then take one employee at random from each firm. This seems problematic to me. Since the employees aren't drawn at random (we can't get two from the same firm), there is no reason for the error terms to be independent. Furthermore, the error terms are likely to be correlated with the explanatory variables, for instance, if different firms reward experience differently. However, Clogg's big point is right. We do know how to extrapolate from a probability sample to the population; that is one thing statisticians are very good at. Regression has a useful role to play in this regard. Population regression equations can be estimated from sample data, with standard errors computed using information about the design – if the investigator really drew a probability sample. Moreover, as Clogg suggests, getting the stochastic assumptions right does not by itself guarantee the causal inferences. That discussion continues in Clogg and Haritou (1994).

[13]For examples where frequentist procedures have the advantage, see Freedman (1963, 1965) or Diaconis and Freedman (1986). Of course, the Bayesians often hold the high ground.

Clogg says that subjectivists can discover objective truth. Of course they can, if they are willing to submit their discoveries to empirical testing. (Objectivists, by the way, should be held to the same standard.) However, he also suggests the possibility of converting subjective priors into objective posteriors by statistical calculations. Here, I disagree. Sensitivity analysis by itself will not do the trick: that several Bayesians would agree among themselves does not make their joint opinion "objective."

To summarize, Clogg and I agree that causal inference by regression modeling is quite problematic; that objectivists and subjectivists will often get quite similar answers, although divergence can be expected too. Furthermore, there is plenty of room for both schools. He suggests that subjective probabilities can be converted into objective ones by statistical calculations; I do not see how such transformations are possible, and will return to this topic shortly.

Henry and I agree that (i) regression equations are most defensible as summaries of data, least defensible as causal (or "behavioral") theories; (ii) the sorts of techniques that work in the physical sciences need not work in the social sciences, because the requisite theory is not in place.

He makes a third and most interesting point: the regression models themselves – which exist in some abstract sphere almost unrelated to any substantive knowledge we might have – tend to supplant behavioral theory. If I am reading him right, his point is illustrated by the sort of sentence found in research abstracts: "The hypothesis tested in this paper is that _____ effects are statistically significant"; the blank is filled out according to the topic of inquiry. Now, as a technical matter, tests of significance apply to hypotheses about a statistical model for data, rather than features of the data set (like $t > 2$). Connecting the data to the model requires theory; but the process that might establish the connection has been short-circuited by the statistical technology.

Let Henry speak for himself (personal communication):

> In journal articles and in courtrooms the regression model itself is proposed as a behavioral model. It is not often challenged as a behavioral model. I believe that it should be interpreted and challenged and defended behaviorally: can one prove that the actions that went into setting salaries followed the 'law' characterized by the regression equation? Do the axioms of an accepted eco-socio-psychological theory lead to an equation like this one?

Clogg and Holland would probably agree, as I do. These are major issues. But Clogg, Henry and Holland go further, suggesting that these issues are more serious than the issues raised by stochastic assumptions. They may well be right; I hardly know which set of flaws is the more fatal.

Henry's comments sadden me in one respect. Statisti ans should be helping to ensure that models make contact with reality; Henry s ggests we are too busy spinning yarns of our own. Part of the problem may ˙ ˙ an intellectual *folie à deux*. According to Lippman, "everybody believes in the [ɪ ɔrmal] law of errors, the experimenters because they think it is a mathematical thec ·əm, the mathematicians because they think it is an experimental fact."[14] Perhap˙ statisticians believe in regression models because they see social scientist fittinç them, while the latter use regression because statisticians keep refining the math matical theory. Henry's solution – "to give up the notion that statistics is an independent discipline" – seems quite drastic, and unlikely to produce the intended effects; erhaps I am too vested in the status quo.

"Priors" or priors? Statisticians can use the Bayesian calculus without being Bayesians. To illustrate the point, I follow Lehmann and consider a random variable X which has a probability density $f_\theta(x)$; the form of f is known, but the density depends on the unknown parameter θ. The statistical problem is to estimate θ. This is a textbook exercise, except that I shall suppose f to be of a novel form, so the estimation problem is new. I shall at times adopt a "pragmatic" position, intermediate between the classical subjectivist and the objectivist, but leaning to the latter. All questions about the validity of the model are set aside for now.

Here are two algorithms for determining estimators.

(i) **Maximum Likelihood.** Estimate θ as the value $\theta_{\mathrm{MLE}}(X)$ which maximizes $f_\theta(X)$; compute the standard error $\mathrm{SE}_{\mathrm{MLE}}(X)$ from the estimated "Fisher information."

(ii) **Bayes.** Choose a probability distribution μ for θ. (This is Lehmann's W.) Using Bayes' rule, compute the conditional distribution $\mu(d\theta \mid X)$ for θ given X. The mean of this conditional distribution is the "Bayes' estimate" $\theta_{\mathrm{Bayes}}(X)$; the square root of the variance is the standard error $\mathrm{SE}_{\mathrm{Bayes}}(X)$.

(This ignores purely technical problems, for instance, the maximum may not exist, or the posterior may not have a finite variance.)

If I pretend to be a classical subjectivist and adopt μ as my prior, then $\mu(d\theta \mid X)$ is my posterior and $\mathrm{SE}_{\mathrm{Bayes}}(X)$ is my uncertainty. But if I do not adopt μ as my prior, the epistemological status of $\mu(d\theta \mid X)$ and $\mathrm{SE}_{\mathrm{Bayes}}(X)$ is obscure. That

---

[14]Cramér (1951, p.232), citing Poincaré (1912).

must be so even if I have chosen μ after much thought and hard work; for instance, computing right Haar measure for a compact non-commutative group or maximizing entropy among some class of distributions.

If I put on my pragmatic but mildly objectivist persona, $\vartheta_{\mathrm{MLE}}(X)$ and $\mathrm{SE}_{\mathrm{MLE}}(X)$ were chosen by one well-known algorithm while $\theta_{\mathrm{Bayes}}(X)$ and $\mathrm{SE}_{\mathrm{Bayes}}(X)$ were chosen by another. The problem is to figure out the operating characteristics, and see which algorithm works better in my situation. The frequentist properties of the Bayes' estimate and the MLE have to be demonstrated; they do not follow automatically.

In textbook examples, both methods work very well and give results that are rather similar. For instance, subject to regularity conditions on the density f,

$$[\theta_{\mathrm{Bayes}}(X) - 2\mathrm{SE}_{\mathrm{Bayes}}(X), \theta_{\mathrm{Bayes}}(X) + 2\mathrm{SE}_{\mathrm{Bayes}}(X)]$$

is asymptotically a 95% confidence interval for θ, providing some comfort to the objectivist. Likewise, the displayed interval will have, asymptotically, 95% probability with respect to the posterior $\mu(d\theta \mid X)$; that is of interest to the subjectivist with prior μ or similar priors, but may not be of interest to the someone with a different prior.

Finally, consider these calculations from the perspective of a committed objectivist. The probability μ was adopted merely as a *ruse de guerre*, taking over part of the Bayesian apparatus but giving it a frequentist spin. There is a price: mathematical work has to be done to establish the frequentist properties of Bayes procedures. Nowadays, this is not as hard as it used to be. (There are two centuries of research on related topics, summarized in Lehmann's comments and in note 12 to my paper.) Even for an objectivist, Bayesian methods can be useful, because the procedures often have frequentist interpretations.

This technical story has a moral. Subjective probabilities were not converted into objective ones. Indeed, there were no subjective probabilities to convert. Calling μ a "prior" does not make it so. This point has particular force with respect to uninformative priors.

Uninformative priors. Many discussants allude to "uninformative priors." Generally, these turn out to be uniform distributions.[15] And in many problems, the appropriate distribution will have infinite mass (for example, Lebesgue measure on the real line). Such uninformative priors do not really quantify subjective degrees of belief. Therefore, neither do the associated posteriors, except by fiat. Uninformative priors are best viewed, at least in my opinion, as devices for generating statistical procedures. These often have good objectivist characteristics, but there

---

[15] More technically, right Haar measure on a locally compact group.

are many exceptions. From the radical subjectivist perspective, on the other hand, statisticians who use uninformative priors risk being incoherent. There is some interesting current research (Eaton and Sudderth, 1993, 1994) on this topic, with new examples of inadmissible and incoherent procedures based on uninformative priors.

Conclusion. My favorite opponent in this debate once made a remarkable concession, not that it interfered with business as usual:

> No sensible social scientist believes any particular specification, coefficient estimate, or standard error. Social science theories ... imply that specifications and parameters constant over situations do not exist .... One searches for qualitative theory ... not for quantitative specifications.[16]

With Hooke's law and the like, we are estimating parameters in specifications that are constant across time – at least to a very good degree of approximation.[17] What are the social scientists doing when they estimate non-existent parameters, and put standard errors on the output? How can that help them search for qualitative theory? Those are among the first of my questions, and I never get answers.

References

Achen, C. (1987), As Statisticians See Us, *Journal of Educational Statistics* **87**, 148-50.

Allen, N. and Holland, P. (1989), Exposing Our Ignorance: The Only 'Solution' to Selection Bias, *Journal of Educational Statistics* **14**, 141-5.

Balke, A. and Pearl, J. (1994), Probabilistic Evaluation of Counterfactual Queries, pp.230-237, Volume I, in the *Proceedings of the Twelfth National Conference on Artificial Intelligence (AAAI-94)*, Seattle, The MIT Press.

Baumrind, D. (1983), Specious Causal Attribution in the Social Sciences: The Reformulated Stepping-Stone Theory of Heroin Use as Exemplar, *Journal of Personality and Social Psychology* **45**, 1289-98.

Blau, P. M. and Duncan, O. D. (1967), *The American Occupational Structure*. New York, Wiley.

Carpenter, K. J. (1981), *Pellagra*. Academic Press.

Cartwright, N. (1983), *How The Laws of Physics Lie*. Oxford University Press.

Clogg, C. C. and Haritou, A. (1994), The Regression Method of Causal Inference and a Dilemma with this Method. Technical report, Department of Sociology,

---

[16] Achen (1987, p.149).

[17] But see Cartwright (1983).

Pennsylvania State University. To appear in V. McKim and S. Turner (eds), *Proceedings of the Notre Dame Conference on Causality in Crisis.*

Cramér, H. (1951), *Mathematical Methods of Statistics.* Princeton University Press.

Daggett, R. S. and Freedman, D. A. (1985), Econometrics and the Law: A Case Study in the Proof of Antitrust Damages, pp. 126-75 in L. LeCam and R. Olshen (eds), *Proceedings of the Berkeley Conference in Honor of Jerzy Neyman and Jack Kiefer*, Vol. I. Belmont, Calif.: Wadsworth.

Dawes, R. (1979), The Robust Beauty of Improper Linear Models in Decision Making, *American Psychologist* **34**, 571-82.

De Leeuw, J. (1985), Review of Books by Long, Everitt, Saris and Stronkhorst, *Psychometrika* **50**, 371-75.

Diaconis, P. and Freedman. D. A. (1986), On the Consistency of Bayes' Estimates, *Annals of Statistics* **14**, 1-87, with discussion.

Dijkstra, T. K. (ed) (1988), On Model Uncertainty and Its Statistical Implications. *Lecture Notes in Econometric and Mathematical Systems*, No. 307, Berlin, Springer.

Eaton, M. L. and Sudderth, W. (1993), Prediction in a Multivariate Normal Setting: Coherence and Incoherence, *Sankhya* **Special Volume 55, Series A, Pt 3**, 481-493.

Eaton, M. L. and Sudderth, W. (1994), The Formal Posterior of a Standard Flat Prior in MANOVA is Incoherent, Technical Report, Department of Statistics, University of Minnesota.

Ehrenberg, A. S. C. and Bound, J. A. (1993), Predictability and Prediction, *Journal of the Royal Statistical Society*, **Series A, Part 2, 156**, 167-206.

Freedman, D. A. (1963), On the Asymptotic Behavior of Bayes Estimates in the Discrete Case, *Annals of Mathematical Statistics* **34**, 1386-1403.

Freedman, D. A. (1965), On the Asymptotic Behavior of Bayes Estimates in the Discrete Case II, *Annals of Mathematical Statistics* **36**, 454-56.

Freedman, D. A. (1985), Statistics and the Scientific Method, pp.343-90 in W. M. Mason and S. E. Fienberg (eds), *Cohort Analysis in Social Research: Beyond the Identification Problem.* New York, Springer.

Freedman, D. A. (1987), As Others See Us: A Case Study in Path Analysis, *Journal of Educational Statistics* **12**, 101-223, with discussion.

Freedman, D. A. (1991), Statistical Models and Shoe Leather, chapter 10 in P. Marsden (ed), *Sociological Methodology 1991*, with discussion.

Freedman, D. A. (1994), From Association to Causation Via Regression. Technical report no. 408, Statistics Department, University of California, Berkeley. To appear in V. McKim and S. Turner (eds), *Proceedings of the Notre Dame Conference on Causality in Crisis.*

Freedman, D. A. and Navidi, W. C. (1989), Multistage Models for Carcinogenesis,

Environmental Health Perspectives **81**, 169-88.

Freedman, D. A. and Navidi, W. C. (1990), Ex-smokers and the Multistage Model for Lung Cancer, *Epidemiology* **1**, 21-29.

Freedman, D. A., Rothenberg, T. and Sutch, R. (1983), On Energy Policy Models, *Journal of Business and Economic Statistics* **1**, 24-36.

Freedman, D. A. and Zeisel, H. (1988), From Mouse to Man: The Quantitative Assessment of Cancer Risks, *Statistical Science* **3**, 3-56 (with discussion).

Freedman, D. A. et al. (1991), *Statistics*, 2nd ed. New York, Norton.

Freedman, D. A. et al. (1993), Adjusting the Census of 1990: The Smoothing Model, *Evaluation Review* **17**, 371-443.

Freedman, D. A. et al. (1994), Adjusting the Census of 1990: Loss Functions, *Evaluation Review* **18**, 243-80.

Holland, P. (1986), Statistics and Causal Inference, *Journal of the American Statistical Association* **81**, 945-70 (with discussion).

Holland, P. (1988), Causal Inference, Path Analysis, and Recursive Structural Equation Models , pp.449-84 in C. Clogg (ed), *Sociological Methodology 1988*. Oxford, Basil Blackwell.

Holland, P. (1994), Probabilistic Causation Without Probability pp.257-92 in P. Humphreys (ed), *Patrick Suppes: Scientific Philosopher*, Vol. 1., Kluwer Academic Publishers.

Hope, K. (1984), *As Others See Us: Schooling and Social Mobility in Scotland and the United States*. Cambridge University Press.

International Agency for Research on Cancer (1986), *Tobacco Smoking*. IARC Monographs on the Evaluation of the Carcinogenic Risk of Chemicals to Humans. Volume 38. Lyon, France.

King, G., Keohane, R. and Verba, S. (1994), *Designing Social Inquiry*. Princeton University Press.

Klein, S. and Freedman, D. A. (1993), Ecological Regression in Voting Rights Cases, *Chance Magazine* **6**, 38-43.

Ling, R. (1983), Review of Correlation and Causation by Kenny, *Journal of the American Statistical Association* **77**, 489-91.

Liu, T. C. (1960), Under-Identification, Structural Estimation, and Forecasting, *Econometrica* **28**, 855-65.

Lucas, R. E. Jr. (1976), Econometric Policy Evaluation: A Critique, pp. 19-64, with discussion, in K. Brunner and A. Meltzer (eds), *The Phillips Curve and Labor Markets*. Vol. 1 of the Carnegie-Rochester Conferences on Public Policy. Supplementary series to the *Journal of Monetary Economics*. Amsterdam: North-Holland.

Luijben, T., Boomsma, A. and Molenaar, I. W. (1988), Modification of Factor Analysis Models in Covariance Structure Analysis: A Monte Carlo Study, pp.70-101 in T. K. Dijkstra (ed), *On Model Uncertainty and its Statistical Implications*. Lecture Notes in Econometric and Mathematical Systems, No. 307, Berlin, Springer.

MacCallum, R. (1986), Specification Searches in Covariance Structure Modeling, *Psychological Bulletin* **100**, 107-120.

Meehl, P. (1954), *Clinical Versus Statistical Prediction: A Theoretical Analysis and a Review of the Evidence*. University of Minnesota Press.

Müller, F. H. (1939), Tabakmissbrauch und Lungcarcinom, *Zeitschrift für Krebsforschung* **49**, 57-84.

Neyman, J. (1923), Sur les applications de la théorie des probabilités aux experiences agricoles: Essai des principes, *Roczniki Nauk Rolniczych* **10**, 1-51, in Polish; English translation by Dabrowska, D. and Speed, T. (1991), *Statistical Science* **5**, 463-80.

Pearl, J. (1994), A Probabilistic Calculus of Actions, pp.454-462 in R. L. de Mantaras and D. Poole (eds), *Uncertainty in Artificial Intelligence*. San Mateo, Calif., Morgan-Kaufmann.

Poincaré, H. (1912), *Calcul des probabilités*. 2nd. ed, Paris.

Robins, J. M. (1986), A New Approach to Causal Inference in Mortality Studies with a Sustained Exposure Period – Application to Control of the Healthy Worker Survivor Effect, *Mathematical Modelling* **7**, 1393-1512.

Robins, J. M. (1987), A Graphical Approach to the Identification and Estimation of Causal Parameters in Mortality Studies with Sustained Exposure Periods, *Journal of Chronic Diseases* **40**, Supplement 2, 139S-161S.

Robins J. M., Blevins, D., Ritter, G. and Wulfsohn, M. (1992), G-Estimation of the Effect of Prophylaxis Therapy for Pneumocytis Carinii Pneumonia on the Survival of AIDS patients, *Epidemiology* **3**, 319-336.

Rogosa, D. (1987), Causal Models do not Support Scientific Conclusions, *Journal of Educational Statistics* **12**, 185-95.

Rogosa, D. (1993), Individual Unit Models versus Structural Equations: Growth Curve Examples, pp.259-81 in K. Haagen, D. Bartholomew and M. Diestler (eds), *Statistical Modeling and Latent Variables*. Amsterdam: Elsevier North Holland.

Semmelweis, Ignac (1861), Atiologie, Begriff und Prophylaxis des Kindbettfiebers. Reprinted in *Medical Classics* **5**, 338-775. There is also a translation by Carter, K. C. (1983), University of Wisconsin Press.

Snow, John (1855), *On the Mode of Communication of Cholera* (2nd ed). London, Churchill. Reprinted in *Snow on Cholera*, 1965, New York, Hafner.

Terris, M. ed. (1964), *Goldberger on Pellagra*. Baton Rouge: Louisiana State University Press.

Veblen, E. (1975), *The Manchester Union-Leader in New Hampshire Elections*.

Hanover, N.H.: University Press of New England.

Foundations of Science
1 (1995/96), 85-97

Diedrik Aerts and Sven Aerts
CLEA, Vrije Universiteit Brussel, Belgium

APPLICATIONS OF QUANTUM STATISTICS IN PSYCHOLOGICAL STUDIES OF DECISION PROCESSES*

Key words: Quantum mechanics, Statistics, Psychology, Quantum structures, Non-Baysian statistics, Interactive statistics.

Abstract. We present a new approach to the old problem of how to incorporate the role of the observer in statistics. We show classical probability theory to be inadequate for this task and take refuge in the epsilon-model, which is the only model known to us capable of handling situations between quantum and classical statistics. An example is worked out and some problems are discussed as to the new viewpoint that emanates from our approach.

1. Introduction.

The quantum probabilility theory is fundamentally different from the classical probability theory. The core of the difference lies in the fact that in quantum probability we are dealing with the *actualization* of a certain property *during the measurement process*, while in classical probability all properties are assumed to have a definite value *before* measurement, and that this value is the outcome of the measurement.

The content of quantum probability is the calculation of the probability of the actualization of one among different potentialities as the *result* of the measurement. Such an effect cannot, in general, be incorporated in a

---

*Supported by the I UAP - III no. 9.

classical statistical framework. A lot has been said about the role of the observer in the human and the social sciences. Indeed, let us consider the most simple example of a psychological questionnaire where the participant is allowed only two possible answers: *yes* or *no*. For certain questions most people will have predefined opinion (eg Are you male/female?, Are you older than 20?). But it is not difficult to imagine a question for which the person who is being questioned has no opinion ready.

For such a situation one answer will be actualized among the two possible answers, at the moment the question is being asked. We are no longer in a classical statistical situation, because the reason why a particular answer is chosen is not so much dependent on the state of the participant or that of the questioner but is in fact highly contextual, that is, the answer is formed at the very moment the interaction between the participant and the examiner is taking place.

Because of the analogue with the quantum mechanical situation, one is very much inclined to investigate whether it is possible to describe such a situation more accurately within a quantum probabilistic framework. Clearly, most questions will not be entirely quantum, nor entirely classical in nature, so what we need is a model that fills the now existing gap between the quantum probabilistic approach and the classical one. Such a model is provided by the epsilon-model that will be given in the next section.

It will be clear that if the introduction of quantum probability theory in psychology is successful, a fundamental problem of psychology, namely the incorporation of the role of the observer in the statistics, can be examined from an entirely new point of view. Similar problems have plagued sociology (in fact, the example we give in section three constitutes a set of sociological problems) and anthropology and we like to think that investigating these areas with the techniques proposed here may lead to important new insights and methodologies.

2. The epsilon-model.

There are several different ways of introducing the epsilon-model and we have chosen the most simple one for the present purpose. (See Aerts D., Durt T., Van Bogaert B., 1992, 1993 and Aerts D., Durt T., 1994.) Consider a particle whose possible states q can be represented by its position on the unit sphere. As a mathematical representation of this state we use a unit vector that we shall call v. Thus we can describe the set $\sum$ of all possible states q_v as follows:

$$\sum = \{q_v \mid v \ lies \ on \ the \ sphere\}$$

Since we want to consider different measurements, we must have some way to identify a certain measurement. We shall do so by introducing another unit vector u. The set M of all measurements $E(u)$ with outcome O is then characterized as follows:

$$M = \{E(u) \mid u \ lies \ on \ the \ sphere\}$$

The outcome O of the measurement $E(u)$ is obtained as follows:

We have a piece of elastic L of length 2. One end of the elastic is attached to the point denoted by the vector u, while the other end is fixed to the diametrically opposite point $-u$.

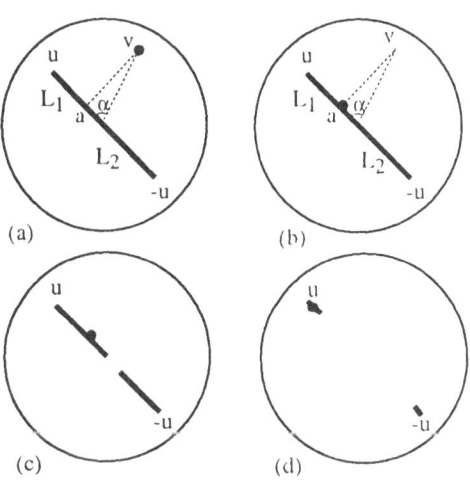

Fig. 1. A representation of the ε-model. A particle located in v (a) falls orthogonally onto an elastic spanned between the two diametrical opposite points u and -u, (b), and sticks to it. Then the elastic breaks, (c), and the particle is pulled to one of the points u, (d), then we say that the measurement has outcome +1, or -u, and then we say that it has outcome -1.

Once this elastic is placed, the particle falls from its original place (denoted by v) onto the elastic, and takes the shortest path when falling, that is, orthogonal to the elastic, and finally sticks on the elastic in a point a.

Now the elastic breaks somewhere. If we consider the two parts of the elastic, the part L_1 from a to $+u$, and the part L_2 from $-u$ to a, then it is obvious that the elastic must break either somewhere in L_1, or else somewhere in L_2.

If it breaks in L_1, the particle in a is drawn towards the point $-u$. Similarly, if the elastic breaks in L_2, the particle in a is drawn towards the point $+u$ In the first case we shall say that the outcome $O = -1$, while in the second case we shall give the experiment the outcome $O = +1$.

This completes the measurement.

We see clearly how the act of measurement transforms the state q_v into the state q_u or q_{-u}, so that our measurement defines a state-transition in an indeterministic way, because we have no knowledge of the point where the elastic will break. The transition probabilities $p(u \mid v)$ and $p(-u \mid v)$ connected to the two possible transitions (q_v to q_u and q_v to q_{-u}) depend on the mechanism by which the elastic breaks.

One can choose different models for this breaking that will lead to different probabilities. We will consider only the most simple case, where the probability of the elastic breaking in a certain segment is proportional to the length of the segment. Let us calculate the transition probabilities for this case.

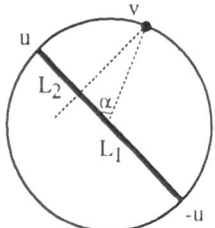

Fig. 2. If we make the hypothesis that the elastic breaks uniformely, the transition probabilities, for the state v changing into u or into -u, is easily calculated.

We see that $p(u \mid v)$ is the probability that the elastic breaks in L_2, and hence equals the length of L_2 divided by 2, which is the length of the elastic. The length of the elastic (see fig. 2) is $1 + a = 1 + u.v = 1 + cos(\alpha)$.

We find the transition probability to be:

$$p(u \mid v) = (1 + cos(\alpha))/2 = cos^2(\alpha/2)$$

Similarly,

$$p(-u \mid v) = sin^2(\alpha/2)$$

These are precisely the quantum mechanical transition probabilities for the spin measurement of a spin $1/2$ particle (eg the electron) in the direction u for a particle with a spin selected (or better: prepared) in the direction v.

What we have here is a precise model for a two-dimensional quantum system.

How are we going to encompass the classical situation in our simple model?

Well, to be sure, it is the breaking of the elastic that invokes the measurement apparatus into the transition probability, so one way to make the transition to a classical regime is simply to limit the length where the elastic can break.

We will call the length where elastic may break $2.\varepsilon$ with $\varepsilon \in [0,1]$, and we will centre this length around the centre of the sphere.

As we can see (Fig. 3) the transition probability will break up in three pieces according to the place where the particle falls. If the particle falls outside the part where the elastic can break, it will evolve deterministically towards the closest fix point of the elastic. If it falls inside the region where the elastic can break, the former reasoning applies: only the length of the elastic is $cos(\alpha) + \varepsilon$ and we have to divide by $2.\varepsilon$ instead of dividing by 2 (which was the old length of the elastic).

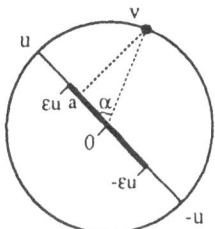

Fig. 3. A representation of the general situation. The elastic connecting the points -u and u, is uniformly breakable between -εu and εu, and unbreakable between - u and -εu, and between εu and u.

We are now in the position to summarize the results:

1. if $cos(\alpha) > \varepsilon$, then $p(u \mid v) = +1$

2. if $-\varepsilon < cos(\alpha) < \varepsilon$, then $p(u \mid v) = (cos(\alpha) + \varepsilon)/(2\varepsilon)$

3. if $cos(\alpha) < -\varepsilon$, then $p(u \mid v) = -1$

3. An example of a decision process

We will now try to show that the epsilon-model is capable of handling situations that are more general than a classical situation and, in fact even more general than a pure quantum statistical situation. In order to do so we must use a questionnaire of three questions. The necessity of using three questions to prove the inadequacy of classical probability theory is the content of the work of L. Accardi (Accardi L., Fedullo A., 1982) and I. Pitowsky (Pitowsky I., 1989). Let us consider then a questionnaire consisting only of the following three questions:

U: Are you a proponent of the use of nuclear energy? (*yes* or *no*)
V: Do you think it would be a good idea to legalize soft-drugs? (*yes* or *no*)
W: Do you think it is better for the people to live in a capitalistic system? (*yes* or *no*)

The reason why we have chosen these questions is related to the fact that we want to isolate groups of people that are strong proponents and opponents for question.

These groups will definitely not change their minds during a simple questionnaire. Yet the questions are also sufficiently complex, so that we can say that a large part of the total examined group (which is, of course, larger than the isolated groups of strong proponents and opponents) did not have an opinion before the question was posed. Although the questions are related, one can easily imagine that every possible combination of answers will be found if a sufficient amount of questionnaires are returned (eg 100 or more).

Let us now make the following somewhat arbitrary, but not a priori impossible assumptions about the probabilities:

We will say that in all cases 50% of the participants have answered question U with *yes*, and that only 15% of the total of persons were convinced of this choice before the question was actually posed. This means that 70% of the participants formed their answer at the moment the question was posed. For simplicity we shall make the same assumptions for questions V and W.

We can picture this situation in the epsilon-model like this:

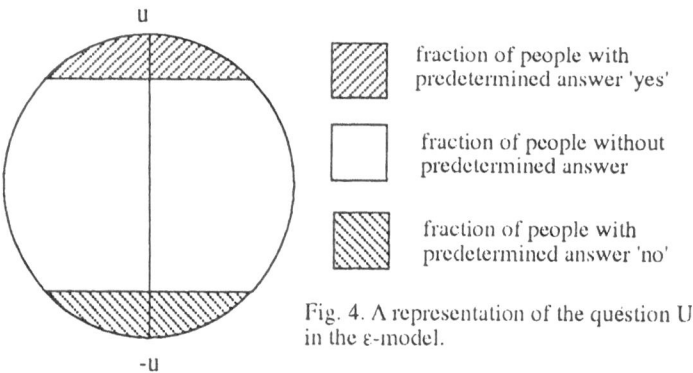

fraction of people with predetermined answer 'yes'

fraction of people without predetermined answer

fraction of people with predetermined answer 'no'

Fig. 4. A representation of the question U in the ε-model.

Next we will have to make some assumptions about the way the three questions are interrelated to one another. This too can be most easily visualized on the sphere of the epsilon-model. We have done this in the following picture for questions U and V:

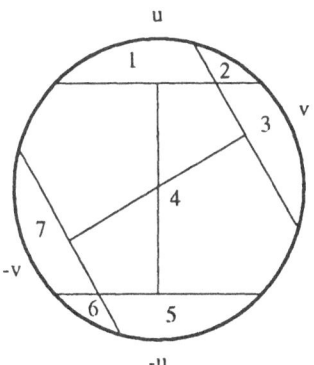

Fig. 5. A representation of the two questions U and V by means of the ε-model. We have numbered the 7 different regions. For example: (1) corresponds to a sample of persons that have predetermined opinion in favour of nuclear energy, but have no predetermined opinion for the other question; (2) corresponds to a sample of persons that have predetermined opinion in favour of nuclear energy and in favour of the legalization of soft drugs; (6) corresponds to a sample of persons that have predetermined opinion against the legalization of soft drugs and against nuclear energy, (4) corresponds to the sample of persons that have no predetermined opinion about none of the two questions, etc...

One can clearly see how a person can be a strong proponent of the use of nuclear energy, while having no significant opinion about the legalization of soft-drugs (this is shaded area nr. II).

Other persons have strong opinions about both questions (the darker shaded areas nr. III and nr. VI). Still others have no prefixed opinion (area nr. VII).

One may ask rightly to what extent these specific assumptions about the relation between the questions represent a real limitation to the use of the epsilon-model in a more general case. We shall postpone this discussion until the next section where we will show how some aspects (eg the apparent symmetry) are only presented here for the sake of simplicity, while others represent problems that need further investigation.

Having said this we will assume the three questions to be related in the following manner:

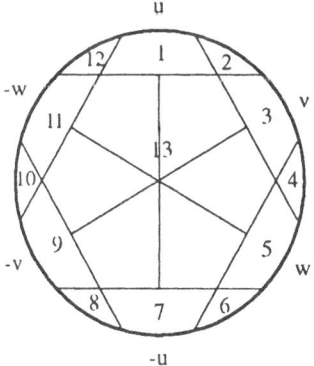

Fig. 6. A representation of the three questions U, V, and W, by means of the ε model. We have numbered the 13 different regions. For example: (1) corresponds to a sample of persons that have predetermined opinion in favour of nuclear energy, but have no predetermined opinion for both other questions; (4) corresponds to a sample of persons that have predetermined opinion in favour of legalization of soft drugs and in favour of capitalism; (10) corresponds to a sample of persons that have predetermined opinion against the legalization of soft drugs and against capitalism, (13) corresponds to the sample of persons that have no predetermined opinion about none of the three questions, etc...

So we could make three pictures like picture 5: one for the relation between questions U and V, one for questions V and W and finally one for questions U and W.

We can see questions V and W are related much in the same way as questions U and V; the only change is a rotation of 60 degrees around the centre of the sphere.

The relation between questions U and W however, is slightly different, as can be seen in the next picture:

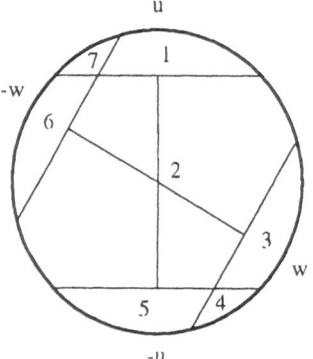

Fig. 7. A representation of the two questions U and W, by means of the ε-model. We have numbered the 7 different regions. For example: (1) corresponds to a sample of persons that have predetermined opinion in favour of nuclear energy, but have no predetermined opinion for the other question; (4) corresponds to a sample of persons that have predetermined opinion in against the legalization of soft drugs and against capitalism; (7) corresponds to a sample of persons that have predetermined opinion for the legalization of soft drugs and for capitalism, etc...

It will be clear that the questionnaire will have a more classical character if the fraction of people that had no opinion before the question (including those people that had an opinion, but changed it because of the questionnaire) is smaller. This fraction is modelled in the epsilon-model by means of epsilon. As indicated in section 2, *epsilon* = 0 means a classical statistical situation, that is a situation where nothing changes as a result of the measurement, while *epsilon* = 1 models a quantum statistical situation (the fraction of non-changing states have measure zero in this case).

In order to make a bridge between these two extremes, we need a concept of probability that involves both notions, i.e. a bridge between a classical conditional probability and a quantum mechanical transition probability. A suitable definition for such a probability is the following:

$p(u = yes \mid v = yes)$ = the probability of somebody answering *yes* to question U, if we can predict that he will answer *yes* to question V.

The important part of this new definition is the word "predict". Hereby we mean the group of strong proponents and opponents previously mentioned. So we have probability that is partly dependent on a reality that exists independent of the observer (...we can predict ...), and partly on a reality that is co-created by the observer.

It is easy to see that in the case of epsilon=1 this definition equals the transition probability, while in the case epsilon=0 the definition is the same as the old definition of the conditional probability. For a more thorough discussion, see - Aerts D., Aerts S., "Interactive Probability Models: From Quantum to Kolmogorovian".

The calculation of this conditional probability is of considerable length and complexity which is the main reason we have chosen not to include it in this article. For the interested reader we refer to - Aerts D., Aerts S., "Interactive Probability Models: From Quantum to Kolmogorovian".

The result of these calculations however, are:

$p(u = yes \mid v = yes) = 0.78$
$p(u = yes|w = yes) = 0.22$
$p(v = yes|w = yes) = 0.78$

The complement of these probabilities are, of course,

$p(u = no|v = yes) = 0.22$
$p(u = no|w = yes) = 0.78$
$p(v = no|w = yes) = 0.22$

We will now prove the following theorem:

Theorem: *The probabilities of the epsilon-model for this epsilon and this choice of u, v and w do not fit in a classical statistical framework.*

Phrased this way, it may seem as if it is only for this particular choice of parameters that the data do not fit a classical statistical model. In fact, most of the values of the epsilon-model do not fit in a classical framework. Mathematically we can extend the following proof which depends on the particular choices made here to a domain of values, by using the continuity of the calculated conditional probability. Let us now give the proof.

Proof:

First of all we have:

$$P(U = yes) = P(U = no) = P(V = yes) = P(V = no) =$$

$$P(W = yes) = P(W = no) = 1/2$$

and,

$$\mu(U_+) = \mu(U_-) = \mu(V_+) = \mu(V_-) = \mu(W_+) = \mu(W_-) = 1/2$$

Where $\mu(U_+)$ denotes the classical probability measure related to finding a *yes* to question U.

If there exists a classical model, then we may use Bayes-axiom to calculate the conditional probability :

$$P(A \mid B) = \mu(A \cap B)/\mu(B)$$

$$1/2.P(V = yes \mid W = yes) = \mu(V_+ \cap W_+) =$$

$$\mu(U_+ \cap V_+ \cap W_+) + \mu(U_- \cap V_+ \cap W_+)$$

$$1/2.P(U = yes \mid W = yes) = \mu(U_+ \cap W_+) = \mu(U_+ \cap V_+ \cap W_+) +$$

$$\mu(U_+ \cap V_- \cap W_+)$$

The right-hand side of the last two equations are found by decomposition over mutually exclusive events.

The above assumptions about the relations between the questions are equivalent to the following choices in the epsilon-model:

$$\varepsilon = \sqrt{2/2},$$

and

$$\alpha_u = 0, \ \alpha_v = \pi/3, \ \alpha_w = 2\pi/3$$

We already mentioned that with these choices we have:
$P(V = yes \mid W = yes) = 0.78$
$P(U = yes \mid W = yes) = 0.22$

Thus we have:

$$\mu(U_+ \cap V_+ \cap W_+) + \mu(U_- \cap V_+ \cap W_+) = 0.39$$

and,

$$\mu(U_+ \cap V_+ \cap W_+) + \mu(U_+ \cap V_- \cap W_+) = 0.11$$

If we subtract the last two equations, we find:

$$\mu(U_- \cap V_+ \cap W_+) = 0.28 + \mu(U_+ \cap V_- \cap W_+)$$

Because of the positive-definite character of the probability measure:

$$\mu(U_- \cap V_+ \cap W_+) \geq 0.28 \tag{3.1}$$

On the other hand we have:

$$1/2.P(U = no \mid V = yes) = \mu(U_- \cap V_+) = \mu(U_- \cap V_+ \cap W_+)+$$

$$\mu(U_- \cap V_+ \cap W_-)$$

Because we also had:

$$P(U = no \mid V = yes) = 0.22$$

We find:
$$\mu(U_- \cap V_+ \cap W_+) \leq 0.11 \tag{3.2}$$

Comparison of equations (3.1) and (3.2) gives us the required contradiction.

We have also proven that this set of data cannot be incorporated in a purely quantum mechanical framework, that is, into a two dimensional Hilbert space. For the interested reader we refer again to (Aerts D., Aerts S., 19XX). These two theorems could have been proven with the set of inequalities that Accardi has provided us with (see Accardi L., 1982) but we thought it would be interesting to present a specific proof.

4. Some problems related to the modelization

In this section we want to investigate the two following questions:

- 1. To what extent can the epsilon-model be enlarged as to encompass some more realistic situations ?

- 2. What does all this mean for future inquiries? Considering the first question, we can point out the following serious restrictions in our example of section 3.

First of all, all data was symmetrical, that is, the chance of the answer "*yes*" was equal to finding the answer "*no*" and this was equal to 1/2 for all three questions. These are only superficial restrictions. Indeed, the epsilon-model has been worked out for the non-symmetrical case as well. The conditional probability for the non symmetrical case was calculated by O. Lévêque.

Secondly, the questions we considered were two dimensional in nature, that is, only *yes* or *no* answers are possible. One could for example say: is not all this too far-fetched; suppose we add a third category saying "maybe", is not all resolved?

The answer is negative. The point is that a person may still change his mind, and whether this be from "*yes*" to "*no*" or from "*no*" to "maybe" or from nothing to "maybe" does not alter the fact that there is a transition as a result of the measurement.

In addition we can reply that the epsilon-model can cope with more dimensional situations just as well, although we have to resort to a more mathematical equivalent so that the possibility of visualizing these situations on a three dimensional sphere is lost.

So, where does the epsilon-model show its weakness?

The weakness lies, of course, in the complexity of the human psychology. People may change their minds in a more or less irreversible way. For some people a conviction is something you carry around the rest of your life, for others it is a mere game that can change from day to day.

This too could be incorporated into the model (in a statistical way of course) by means of another parameter that parametrizes that amount of induced change. We have called this parameter η, and although some preliminary work has been done, a whole area of investigation remains unexplored.

What does this all mean for future questionnaires?

Well, first of all we need to be fully aware of these facts. We can try two strategies: Firstly we may try to make these transitions relatively unimportant. We can do this by asking unimportant easy answers, but this is not a very satisfactory way of resolving the problem. Secondly we can try to create an atmosphere wherein a person is less likely to change his mind: an impersonal questionnaire by a machine instead of an interview with a nice looking lady.

All this has already been tried.

Apart from this, we suggest another approach: investigation of the changing. Indeed, our mind was once a tabula rasa, and our very ideas and convictions are all formed as the result of these interactive changes. In order to understand better these interactive processes we will have to see humanity as a whole and not just as an ensemble of individuals. We hope that this model, even in its present rather limited form, can help in this kind of investigation.

5. Conclusions

Although much work needs to be done we have shown that the inter-active transitions which take place in human interactions, are unlikely to be understood by means of classical probability theory. What we need is an interactive form of statistics, somewhere between quantum and classical probability. We have shown a first example of such a model and discussed some future routes to be followed.

References

Aerts, D., Durt, T., Van Bogaert, B. (1992), A physical example of quantum fuzzy sets and the classical limit, *Tatra Mountain*, Math. publ., vol 1, 5-14.

Aerts, D., Durt, T., Van Bogaert, B. (1993), Indeterminism, non-locality and the classical limit, in: *Symposium on the Foundations of modern Physics*, Helsinki, World Scientific, Singapore.

Aerts, D. (1993), Quantum structures due to fluctuations of the measure-ment, *International Journal of Theoretical Physics*, vol 32, nr.**12**.

Aerts, D., Durt, T. (1994), Quantum, Classical and Intermediate: an illus-trative example, in: *Foundations of Physics*.

Accardi, L. Fedullo, A. (1982), On the statistical meaning of Complex num-bers in Quantum Mechanics, *Al nuovo cimento*, Vol.34, nr.**7**.

Pitowsky, I. (1989), Quantum Probability – Quantum Logic, Springer Verlag.

Aerts, D., Aerts S., Interactive Probability Models: From Quantum to Kol-mogorovian, preprint, VUB pleinlaan 2, 1050 Brussels, Belgium

Foundations of Science
1, 99-118, 1995/96

Maria Carla Galavotti
Department of Philosophy
University of Trieste, Italy

OPERATIONISM, PROBABILITY AND QUANTUM MECHANICS

Key Words: Operationism, Quantum mechanics, Frequentism, Subjectivism.

Abstract. This paper investigates the kind of empiricism combined with an operationalist perspective that, in the first decades of our Century, gave rise to a turning point in theoretical physics and in probability theory. While quantum mechanics was taking shape, the "classical" (Laplacian) interpretation of probability gave way to two divergent perspectives: frequentism and subjectivism. Frequentism gained wide acceptance among theoretical physicists. Subjectivism, on the other hand, was never held to be a serious candidate for application to physical theories, despite the fact that its philosophical background strongly resembles that underlying quantum mechanics, at least according to the Copenhagen interpretation. The reasons for this are explored.

In the same years in which logical empiricism was taking shape as a philosophical doctrine, notably in the Twenties, the empiricist outlook, combined with an operationalist perspective, led to a turning point in theoretical physics and in probability theory.

The advent of quantum mechanics in the years 1925-27, at least as far as the matrix mechanics of Werner Heisenberg and Max Born is concerned, was in fact brought about following (1) the radically empiricist principle that unobservables should be banished from physics, and (2) the operationalist maxim according to which physics can only deal with measurable quantities.

By that time, probability had been emancipated from the strictures of the classical – Laplacian – interpretation. This gave way to the frequentist notion of probability, formulated by Richard von Mises, again in a purely empiricist and operationalist fashion. Frequentism was soon widely accepted both by scientists and logical empiricist philosophers, and became the "official" interpretation of probability in science.

At around the same time, another view of probability was put forward in the same empiricist and operationalist spirit that we find at the root of frequentism. This is the subjective interpretation of probability, due to Frank P. Ramsey and Bruno de Finetti. Unlike frequentism, the subjective notion of probability had to wait more than twenty years to gain widespread acceptance. Even when subjectivism became popular in the Fifties, after L.J. Savage's work, it found its main applications within statistics, economics, game theory, decision theory, and the like. As a matter of fact, the subjective interpretation of probability has never been taken seriously either by logical positivists, or by physicists who still today seem to think that the whole issue of the interpretation of probability revolves around the contraposition between frequentism and the Laplacian "classical" interpretation. The reasons why this should be so are worth exploring, as the philosophical position underlying the subjective notion of probability bears a substantial resemblance to that underlying quantum mechanics, at least according to the so called Copenhagen interpretation. My attempt to clarify this point will explore some aspects of empiricism in our Century that are often neglected, namely those connected to the foundations of probability, particularly in relation to subjectivism, and quantum mechanics.

The investigation will start with a sketchy recollection of the main ideas characterizing the statistical notion of probability that found its main expression in von Mises' frequentist interpretation, meant as a purely empiricist view of probability, built in an entirely operationalist fashion. We will then see how the same operationalistic attitude is to be found in the work of the upholders of the Copenhagen interpretation of quantum mechanics, as well as in de Finetti's view of probability. There will follow a discussion of the main similarities and dissimilarities between the subjective notion of probability and the philosophy of quantum mechanics.

1. From classical to frequentist probability.

Since the time of P.S. de Laplace (1814), what is known as the "classical" interpretation of probability long represented a prominent way of looking at probability. According to this interpretation, probability expresses our ignorance about the true course of events, and can be determined as the ratio of favourable to all the equally possible alternatives, on the basis of the so called "principle of insufficient reason".

Within Laplace's perspective, this notion of probability goes hand in hand with a deterministic world view, reflecting the mechanistic character of science pertaining to classical mechanics. Accordingly, Laplace's idea of an "omniscient being" who would be able to predict all future events with certainty has been seen as "the most extreme formulation of Newton's determinism".[1]

The Laplacian notion of probability has been labelled as "subjectivism" because it conceives probability as relative to human knowledge and indeed as the expression of the human impossibility of gaining complete knowledge of what is going on in our world. In view of the subsequent formation of a subjectivist interpretation of probability in this century, it is preferable to talk of a "classical" view of probability in connection with the Laplacian view which is certainly epistemic.

By 1900 the classical notion of probability had gradually become out-dated, and a statistical view, according to which probability is strictly connected to frequencies, became prominent. One important factor in this passage to a new conception of probability was the new trend taken by physics. According to R. von Mises, who developed the most prominent version of the frequentist interpretation of probability, the turning point was reached when Boltzmann "conceived the remarkable idea of giving a statistical interpretation to one of the most important propositions of theoretical physics, the Second Law of Thermodynamics".[2] As a matter of fact, the study of microphenomena gave rise to a number of theories where probability enters as an essential ingredient, like the kinetic theory of gases, Brownian motion and the theory of radioactivity. In the hands of R. von Mises, and later H. Reichenbach, frequentism became the "official" interpretation of probability, to be applied to phenomena described by such theories. Statistical probability corresponded to a widespread scientific practice, and became for this reason most popular with physicists, who accepted it quite unproblematically.

The frequentist notion of probability also pervaded the mathematical theory of probability, where around the period 1900 - 1930 laws of large numbers and other limiting theorems about the behaviour of relative frequency were derived on the assumption of probabilistic independence or analogous weaker properties. In the same period theoretical statistics grew enormously and various methods for testing statistical models and estimating statistical parameters were developed. With very few exceptions – notably that of H.

---

[1] Von Mises 1928, p. 176.

[2] *ibidem*, p. 174.

Jeffreys – frequentism was also accepted within this branch of statistics. To sum up, one can say that frequentism superseded the classical interpretation and became the "received view" of probability.

2. Statistical probability.

The most perspicuous version of the frequentist interpretation of probability was worked out by R. von Mises. In his *Probability, Statistics and Truth* (German original 1928) he devotes one chapter to "Statistical problems in physics". There he argues in favour of a view of probability taken as relative frequency and defined on the basis of the fundamental notion of "collective". Then he argues that probability can be applied to physical phenomena insofar as these are reducible to "chance mechanisms" having the features of a collective.

A collective is defined by means of two fundamental conditions: firstly, the relative frequencies of its attributes must possess limiting values; secondly, these limiting values must remain the same in all partial sequences which may be selected from the original one in an arbitrary fashion (in other words, the sequence is random). According to von Mises "it is possible to speak about probabilities only in reference to a properly defined collective".[3] The principles of the probability calculus are defined on the basis of four different ways in which one can "derive new collectives from others". The four "fundamental operations" of selection, mixing, partition and combination are devised for that purpose. While giving a way of measuring probabilities on the basis of frequencies, von Mises' approach is meant as a purely empirical approach to probability. Probability is operationally reduced to a measurable quantity.

One can object to the operationalist character of von Mises' theory, on the basis that it uses infinite sequences. But von Mises thinks that that is the only way to make a mathematically accurate probability theory. Probability as an idealized limit can be compared to other limiting notions such as velocity or density.[4] Von Mises is not very precise about methodological questions, but, by and large, the application of probability proceeds as follows. First, the hypothesis is made that a repeatable event ("mass phenomenon") displays stability of relative frequency and randomness. Deductions are made in the mathematical theory. The consequences are tested against experience, through retranslation from the mathematical theory. Von Mises did not think one could formulate general rules of retranslation from theory to experience. Instead, the criteria of application are contained in the inductive

---

[3] *ibidem*, p. 28.
[4] See von Mises 1951, p. 168.

process that leads from ordinary experience to an exact theory. In the case of probability, the inductive process consists in the recognition of stability of relative frequency and of insensitivity to selection of subsequences.[5]

Once an inductive process has led to an exact theory, it is treated in the hypothetico-deductive fashion. At all costs von Mises wants to avoid the idea of any inductive relation of a logical type between a finished theory and the observable world, for this would bring an epistemic component to scientific inference. We can barely discern such a component in von Mises' admission that with finite sequences, "our theory shows what can be expected, on the average ...".[6] Further on he says that "the notion of the infinite collective can be applied to finite sequences of observations in a way which is not logically definable, but is nevertheless sufficiently exact in practice".[7]

When one comes to physical phenomena, as previously mentioned according to von Mises, probability can be applied to such phenomena only insofar as they can be reduced to "chance mechanisms" having the features of collectives. Von Mises mentions the following as typical parts of physics that can be treated probabilistically: 1. the kinetic theory of gases, 2. Brownian motion, 3. radioactivity, 4. Planck's theory of black-body radiation. After having shown how the phenomena which are the object of the above theories can be reduced to collectives and treated probabilistically, he discusses "the new quantum statistics created by de Broglie, Schrödinger, Heisenberg and Born".[8] He seems to think that frequentist probability can be extended to this new field without many problems. Finally, von Mises discusses Heisenberg's uncertainty principle and welcomes it as a possible basis on which the old and the new physics could be unified on probabilistic grounds. This testifies to an indeterministic attitude, that – as we shall see – von Mises shared with those who created quantum mechanics. In fact the advent of quantum mechanics, through Heisenberg's uncertainty principle, brought with it the idea that indeterminism could be a correct hypothesis about the world, at least as respectable as Laplace's determinism.

When quantum mechanics took shape, namely around 1925-27 with the work of Werner Heisenberg, Niels Bohr and Max Born [9], it did not destroy the frequentist construction, despite the fact that it contained elements that

---

[5] *ibidem*, pp. 142-144.

[6] Von Mises 1928, p. 84.

[7] *ibidem*, p. 85.

[8] *ibidem*, p. 211.

[9] See Pais 1982, p. 1194, where these three men are indicated as those who brought to an end that revolutionary period of formation of quantum theory that was started in 1900 by Planck and carried on by many others, including A. Einstein, N. Bohr, L. de Broglie, S.N. Bose, E. Schrödinger.

appear to be in contrast with frequentism. The main problem was clearly related to the single case. The tendency on the part of physicists was to overcome the problem by simply extending the frequentist interpretation to the single case.

3. Probability in physics.

An operationistic attitude analogous to von Mises' is to be found in the work of physicists. Einstein in 1916 introduced a probability distribution for the energy levels e_1, e_2, e_3,...of atoms, as well as transition probabilities between these levels. In a jump to a lower level, e_m to e_n, energy is emitted in the form of light of a frequency (colour) ν determined by Bohr's frequency condition $h\nu = e_m - e_n$. The transition probabilities correspond to the light intensities of the respective frequencies. With a great many atoms and transitions, they are easily measured, offering an extreme illustration of the idea that probabilities are observable quantities, which inspires von Mises' conception.

The same operationalistic attitude characterizes the writings of those who created the Copenhagen interpretation of quantum mechanics. Let me recall some of the leading ideas at work here. In his paper "Statistical interpretation of quantum mechanics" Born recollects the process that brought about quantum mechanics – referring to the famous paper by Heisenberg of 1925 which contains the principles that were to be developed as matrix mechanics – and credits Heisenberg (his assistant at the time) with having made the decisive turn in the development in quantum mechanics. According to Born, such an accomplishment was due to the fact that Heisenberg "cut the Gordian knot by a philosophical principle and replaced guesswork by a mathematical rule".[10] While the mathematical rule refers to Heisenberg's calculation of transition amplitudes, later to be developed by Born himself, the philosophical principle "asserts that concepts and pictures that do not correspond to physically observable facts should not be used in theoretical description".[11] In other words, the principle says that only measurable quantities can be the object of physics, and physics should limit itself to observables.[12] By following this maxim of conduct Heisenberg "banished the picture of electron orbits with definite radii and periods of rotation, be-

---

[10] Born 1969, p. 91.

[11] *ibidem.*

[12] In the paper recollected by Born, Heisenberg says that his method is "to determine quantum-theoretical data using relations between observable quantities" (1925, p. 276). The maxim is neatly expressed also in Heisenberg 1927, where it is said that "Physics ought to describe only the correlation of observations" (p. 82).

cause these quantities were not observable".[13] While doing so, he opened the way to the idea that transition probabilities should replace the description of motion through spatial coordinates expressed as functions of time. The maxim according to which no unobservables should be admitted in the realm of physics is for Born a powerful and innovative heuristic principle, that not only made possible the turning point in quantum mechanics, but also played a crucial role in the theory of relativity. In Born's words: "When Einstein, in setting up his theory of relativity, eliminated the concepts of the absolute velocity of a body and of the absolute simultaneity of two events at different places, he was following the same principle".[14]

In addition to working as a heuristic principle, the maxim gives us the key to the interpretation of quantum theory held by the Copenhagen school. It is through this maxim that one should read Heisenberg's claim that there is only one answer to the question: "What happens 'really' in an atomic event?", and this amounts to saying that "the term 'happens' is restricted to observation".[15] This represents a key feature of the Copenhagen interpretation of quantum theory, responsible for the endless discussion on the EPR paradox, Schrödinger's cat argument, and more generally the debate connected with the so-called incompleteness of quantum mechanics. Here we leave aside that debate and concentrate on the kind of empiricism that characterized the birth of quantum mechanics. The principle that the theory is restricted to observables immediately leads to the notion of measurement, since the observables themselves are nothing but the objects of measurements. This, in turn, is strictly connected with the probabilistic character of quantum mechanics, for "what one deduces from an observation is a probability function".[16] It is worth recalling how according to Heisenberg a probability function resulting from an observation (like initial position and velocity of an electron in a cloud chamber) "represents a mixture of two things, partly a fact and partly our knowledge of a fact".[17] More precisely, "The probability function combines objective and subjective elements. It contains statements about possibilities or, better, tendencies ('potentia' in Aristotelian philosophy), and these statements are completely objective, they do not depend on any observer; and it contains statements about our knowledge of the system, which of course are subjective in so far as they may be different for different observers. In ideal cases the subjective element in the probability function may be practically negligible as compared with the

---

[13]*ibidem.*

[14]*ibidem.*

[15]Heisenberg 1958, p. 40.

[16]*ibidem*, p.34.

[17]*ibidem*, p. 53.

objective one The physicists then speak of a 'pure case'".[18] The interesting aspect of Heisenberg's position – quite apart from the somewhat obscure reference to the Aristotelian notion of "potentia", that has captured the attention of many readers [19] – is the explicit recognition that the "objective elements" entering in the probability function are determined by the theory. In other words, they follow as consequences from the laws of quantum physics.

Apart from the ideal case – the pure state, to which we will return later – the objective elements are always combined with the subjective elements connected to observation. Observation, i.e. measurement, contains two elements of uncertainty: the disturbance connected to all measurements plus the interaction with the measuring device, which is the object of Heisenberg's uncertainty relation. Such an interaction represents a peculiar aspect of quantum mechanics, that leads to a probability distribution over the next observation to be performed. The next observation "selects of all possible events the actual one that has taken place",[20] and therefore the transition from the "possible" to the "actual", "takes place during the act of observation".[21] Now, the absence of precise knowledge of what happens between two observations is responsible for the fact that in quantum mechanics "the word 'happens' can apply only to the observation, not to the state of affairs between two observations".[22] Here lies the foundation for the maxim that only observables should be admitted in quantum mechanics.

The peculiar features of "quantum objects" and the inevitable interaction connected to their observation seems to introduce a subjective element in the description of atomic events, and consequently in the kind of empiricism upheld by the Copenhagen interpretation. As Heisenberg puts it: "the measuring device has been constructed by the observer, and we have to remember that what we observe is not nature in itself but nature exposed to our method of questioning".[23] However, it should be emphasized that the form of empiricism we see here does not involve a subjectivistic attitude ultimately based on sensations. Heisenberg makes repeated claims to the effect that his own interpretation of quantum mechanics does not introduce "the mind of the physicist as part of the atomic event".[24] When an observation is made and an interaction with the measuring device takes place, this

---

[18] *ibidem*, p.41.

[19] See for example Margenau and Park 1967.

[20] Heisenberg 1958, p. 42.

[21] *ibidem*.

[22] *ibidem*.

[23] *ibidem*, p.46.

[24] Heisenberg 1958, p. 43.

process involves "the physical not the psychical act of observation" and "it is not connected with the act of registraticn of the res ... by the mind of the observer".[25] Accordingly, for Heisenberg "the sensual perception of the observer" should not be included in the realm of scientific theories and it is to be clearly recognized that "the Copenhage: interpretation of QT is in no way positivistic".[26]

A similar attitude is taken by Born, who discusses many times in his writings what he calls "empirical realism", of which the task is "tc state as clearly as possible the nature of the reality which forms the subject of natural science".[27] In this connection, he maintains that "it is not the .eality of sense-perceptions, of sensations, feelings, ideas, or in short the subje:tive and therefore absolute reality of experience. It is the reality of things, of objects, which form the substratum underlying perception".[28] And he identifies the only criterion for ascribing a meaning to the formalism of quantum mechanics with the exhibition of its "relation with the observational concepts of the experimenter".[29]

Here the word "realism" should not lead us astray, for Born seems to be as much of an anti-realist as Heisenberg. " ...scientific forecasts – Born says – do not refer directly to 'reality' but to our knowledge of reality".[30] A passage by Heisenberg puts it even more clearly: "human language permits the construction of sentences which do not involve any consequences and which therefore have no content at all – in spite of the fact that these sentences produce some kind of picture in our imagination; e.g., the statement that besides our world there exists another world ...One shou.d be especially careful in using the words 'reality', 'actually', etc., since these words very often lead to statements of the type just mentioned".[31]

As we have seen, Born's empiricism, like Heisenberg's, is not the kind based on sense-perception, being based rather on the results of observations, or in other words on measurement. Born also emphasizes the importance of laws. In defining the object of science he refers to "the accordance with general laws we detect in phenomena".[32] In other words, science is made of laws, and both the meaningfulness of scientific talk and our ability to make predictions rest on them. "...the so-called 'laws of nature' allow us

---

[25] *ibidem*, pp. 42-43.

[26] *ibidem*, p. 133.

[27] Born 1969, p.16.

[28] *ibidem*. See also Born 1953, where it is said that "it is psychologically and physiologically wrong to regard the crude sense impressions as the primary data" (p. 147).

[29] *ibidem*, p.27.

[30] Born 1969, p.163.

[31] Heisenberg 1930, p. 15.

[32] *ibidem*, p. 16.

to draw conclusions from our limited, approximate knowledge at the moment on a future situation which, of course, can also be only approximately described".[33]

This brings us back to the probabilistic character of quantum mechanics, a character that for Born can be extended to all of science, including classical mechanics. In his paper "Statistical Interpretation of Quantum Mechanics" Born claims determinism is nothing but "an article of faith" created by the enormous success of Newtonian mechanics. He goes on to point out that "determinism becomes complete indeterminism if one admits even the smallest inaccuracy" in the results of measurements, and speaks in favour of a statistical formulation of ordinary mechanics. This would be in agreement with the "heuristic principle employed by Einstein in the theory of relativity and by Heisenberg in quantum theory" according to which "concepts that correspond to no conceivable observation ought to be eliminated from physics".[34] Within such a reformulation the probability distribution of values should be substituted for precise values of physical quantities. This would definitely show that determinism "is an idol, not an ideal" in the investigation of nature. Incidentally, it is worth noting that Born praises von Mises for having said that "even in classical mechanics, determinism is only a fiction and of no practical significance".[35] We have seen at the end of last section how according to von Mises there is a continuity between classical and quantum mechanics.

The question now arises of what conception of probability was embraced by the upholders of the Copenhagen interpretation of quantum mechanics. However, the question does not seem to have been the object of much attention on their part. Frequentism seems to be tacitly assumed, and references are to von Mises' work. This is not surprising, in view of the empirical character of von Mises' conception, whose operationism probably made it quite palatable to people like Born and Heisenberg. However, within quantum mechanics the attribution of probabilities to the single case is generally admitted. Born claims that the statistical interpretation of quantum mechanics is meant to apply to the formalism of quantum mechanics "in any individual case"[36], and Heisenberg wants to talk of "the observation of 'the electron in a single atom'".[37]

4. Modern subjective probability

[33] *ibidem*, p. 163.

[34] *ibidem*, p. 97.

[35] See Born 1969, p. 28 footnote, where reference is made to von Mises 1928.

[36] *ibidem*, p. 27.

[37] Heisenberg 1958, p.64.

Almost in the same years when frequentism took shape with von Mises and quantum mechanics became a fully fledged physical theory, F.P. Ramsey and B. de Finetti [38] were proposing their subjective theories of probability. These modern theories are far wider in their range of application than the classical Laplacian one, and do not presuppose any a priori limitations to the form of probability laws. Probability is the expression of an attitude, called "degree of belief", that a person takes towards an event. It can be any event, past, present or future, provided the person is uncertain about its occurrence. No stand needs to be taken as to the reason for the uncertainty, be it mere ignorance, indeterministic chance, complexity of causes, and so on.

In the modern conception, subjective probabilities are quantitative expressions of degrees of belief. Any assignment of probability is admissible, provided it is consistent in a specific sense. This criterion, often called coherence, can be expressed as follows. Subjective probabilities are to be assigned so that, if they are used as the basis of betting ratios, they should not lead to a sure loss. This requirement ensures that subjective probabilities obey the usual rules of probability calculus.

In further elaborations of the theory, utility values are assigned to the different events. If the person's actions affect the occurrence of events, the rule is to choose that action which maximizes the expected utility, where expectation is relative to the probability assignment made by the person. From an epistemological point of view, a person's subjective probabilities can be compared to his beliefs more generally.

Ramsey's and de Finetti's main accomplishments are quite well known [39], so, instead of recalling them, I will put forward a few remarks on a paper written by de Finetti in 1978, called "Einstein, Originality and Intuition". To my knowledge, this is the only paper de Finetti devoted to the character of physical problems, though in various places he touches upon related issues, like indeterminism. In this paper de Finetti takes the occasion of the forthcoming centenary of Einstein's birth (1879) to comment on Born's book *Physics in my Generation*, a choice he motivates on the basis of the affinity between Born's ideas and his own.

That de Finetti must have found Born's ideas congenial can hardly be surprising, if one considers a passage like the following: "I think that ideas such as absolute certainty, absolute precision, final truth, etc. are phantoms which should be excluded from science. From the restricted knowledge of

---

[38] See F.P. Ramsey's paper "Truth and Probability", written in 1926 and published posthumously in Ramsey 1931, and de Finetti 1929, 1931a, 1931b, 1931c. More references to de Finetti will be found in the following pages.

[39] The reader is referred to Galavotti 1989 and 1991 for further discussion on the topic.

the present state of a system one can, with the help of a theory, deduce conjectures and expectations of a future situation, expressed in terms of probability ... This loosening of the rules of thinking seems to me the greatest blessing which modern science has given us. For the belief that there is only one truth and that oneself is in possession of it, seems to me the deepest root of all that is evil in the world".[40] The passage above is taken from Born's paper "Symbol and Reality", but not knowing that, one could equally well take it to be from one of the many papers by de Finetti, for instance "Probabilismo".[41] It reflects not only the position that de Finetti called "probabilism", but identifies its origin in the refusal of the notion of truth, exactly like de Finetti in his paper of 1931.

Turning to de Finetti's discussion of Einstein's views on space and time, we encounter the claim that he finds such views congenial for "two specific reasons", the first being that they were based "on an *operative* concept: the experimentation with light signals" and the second lying in their relativistic character. He goes on to qualify Einstein's perspective as "decidedly empiricist and pragmatist" like his own, and praises Born for his criticism of determinism and causality.

Indeed, operationism is as much a characteristic of the authors mentioned in the preceding sections as of de Finetti's conception of probability. This is rooted in the conviction that "for any proposed interpretation of probability, a proper operationalist definition must be worked out: that is, an apt device to measure it must be constructed".[42] The essential ingredients of the operationalist definition of probability as "degree of belief" are the previously mentioned notions of "betting quotients" and "coherence". Such a definition confers applicability to probability, which in itself is a primitive, psychologistic notion expressing the "psychological sensation of an individual".[43]

Operationism is for de Finetti a criterion of adequacy, in the light of which subjectivism turns out to be the only acceptable interpretation of probability. He claims that this is the only one amenable to an operative definition. On such grounds de Finetti criticizes all the other interpretations of probability: classical, logical and frequentist alike. In particular, he claims that such interpretations take a "rigid" approach to probability, because they assume the existence of objective reasons for choosing a particular probability function among all "admissible" ones. In so doing, they confuse logical with extra-logical considerations in the assessment of probability, and appeal to extra-logical elements, which to him are merely the

---

[40] Born 1969, p. 143.
[41] See de Finetti 1931a.
[42] De Finetti 1972, p. 23.
[43] De Finetti 1931b, p. 302.

product of a "metaphysical mentality". While refusing all the other perspectives when taken as interpretations of probability, de Finetti claims that they can instead give practical criteria for the evaluation of probability in specific situations.

It is interesting to recall the essential criticisms moved by de Finetti against frequentism.[44] His objections are twofold: in the first place, he claims that the pretence to define probability by means of frequency brings a contradiction with it, because the link between probability and frequency is not to be taken as given, being rather a matter for demonstration. As a matter of fact, demonstrating the relation between probability and frequency is a main task of probability calculus, but this becomes impossible if such a relation is postulated. De Finetti admits that frequency and probability are strictly connected, and maintains that the evaluation of frequencies enters into probability judgments, in a way that is reflected by his "representation theorem" (on which I will add something later). However, for de Finetti the evaluation of probabilities is a complex judgment that involves a number of context-dependent elements, in addition to frequencies. Such additional ingredients of probability judgments are peculiarly subjective, and this is de Finetti's second objection against frequentism.

While inadequate as an interpretation of probability, within a subjectivist perspective frequentism can be taken as a practical criterion for evaluating probabilities: namely in cases where we judge certain events to show similarities that allow us to put them in one sequence of repetitions. This judgment, however, is subjective. It is worth noting that de Finetti claims one cannot talk of repetitive events in the objective sense, but only of single events.

Considerations similar to the above also apply to the "classical" interpretation, which is dismissed as a way of defining probability, but is retained as a criterion that suggests choosing a uniform distribution of initial probabilities, as is done in well known cases such as games of chance. However, "the belief that the a priori probabilities are distributed uniformly is a well defined opinion and is just as specific as the belief that these probabilities are distributed in any other perfectly specified manner".[45]

Going back to de Finetti's paper on Einstein, the author finds points of analogy between his approach to probability and Born's ideas. Thus, Born suggests one ought "to reformulate classical mechanics in such a way that it dealt only with not sharply defined states".[46] The idea of reformulating classical mechanics in probabilistic terms is clearly in line both with de Finetti's

---

[44] See especially de Finetti 1936 and 1941.

[45] De Finetti 1951, p. 222.

[46] Born 1969, p. 163, quoted by de Finetti 1978, p. 127.

probabilism and his indeterministic leaning.[47] What is perhaps even more interesting in this context is a remark by de Finetti pertaining to Born's further suggestion that a reformulation with unsharp states would refer in quantum mechanics "to the notion of 'mixture of pure cases'". De Finetti points out that the quantum theoretical notion of "mixture of pure cases" "coincides with the one deriving from my definition of *exchanceability*".[48] He goes ón to recall how the process of assigning probabil· ies according to the criterion of exchangeability allows learning from expe.ience, unlike what happens if independence is assumed, and showing how exchangeable probability assignments vary with observed frequencies in such a way as to tend to an asymptotical agreement.

Before we return to the analogy between the notion of a mixed state and the "representation theorem" for exchangeable probabilities, it is in order to say something about the latter.[49] Within de Finetti's perspective, the representa:i:n theorem serves two purposes: firstly, it makes possible the so-callec' reduction of objective to subjective probability; secondly, it assures scientif.c applicability of the subjective notion of probability. The representati~n theorem is based on the notion of exchangeability. Events belonging to a sequence are *exchangeable* if the probability of h successes in n events is the same, for whatever permutation of the n events, and for every n and $h \leq n$. The representation theorem says that the probability of exchangeable events can'be represented as follows: imagine the events were probabilistically independent, with a common probability of occurrence p. Then the probability of a sequence with h occurrences in n would be $p^h(1-p)^{n-h}$. But if the events are only exchangeable, the sequence has a probability $\omega_h^{(n)}$, representable according to de Finetti's representation theorem as a *mixture* over the $p^h(1-p)^{n-h}$ with varying values of p:

$$\omega_h^{(n)} = \int p^h(1-p)^{n-h} f(p) dp.$$

Here $f(p)$ is a uniquely defined density for the variable p, or in other words, it gives the weights $f(p)$ for the various values $p^h(1-p)^{n-h}$ in the above mixture.

De Finetti's notion of exchangeability means that the locations of the successes make no difference in the probability of a sequence. These need not

---

[47]See for example de Finetti 1931c, where a stochastic term is added to the equations of classical motion.

[48]De Finetti 1978, p. 126.

[49]The result known as the "representation theorem" was obtained by de Finetti as early as in 1928 (see de Finetti 1929), though its best known formulation is contained in de Finetti 1937.

be independent sequences. An objectivist who wanted to explain subjective probability, would say that the weighted averages are precisely the subjective probabilities. But de Finetti proceeds in exactly the opposite direction, by saying that *starting* from the subjective judgment of exchangeability, one can show that there is only one way of giving weights to the possible values of the unknown objective probabilities. We can now see how the reduction of objective to subjective probability is made. Objective probabilities are a construct one has no need for if one follows de Finetti. Furthermore, exchangeability is a more general notion than probabilistic independence, and allows the subjective probabilities and observed frequencies to approach each other. One obtains this by conditionalization on observations. In other words, the representation theorem shows how information about frequencies interacts with subjective factors within statistical inferences. This is how the task of giving applicability to subjective probability is accomplished.

Going back to the analogy with the notion of a mixed state, this is indeed striking with respect to such authors as Born and Heisenberg, especially in view of Heisenberg's claim – recalled above – to the effect that there are two components of a probability function in quantum mechanics: a subjectivist and an objectivist one. As a matter of fact, in subsequent literature it has been shown that the analogy does not go all the way through, because there is no unique decomposability of mixed states into pure states in quantum mechanics, so that the latter states are not in general reducible to the former. The interpretation of quantum states is a matter of much debate, and we cannot go into that here.[50]

5. Conclusion.

The main question of this paper can now be put as: Why was subjectivism never taken seriously either by logical positivists or by physicists, despite its empirical and operationalist character?

After all, there seem to be remarkable philosophical similarities between de Finetti's subjectivism and the Copenhagen interpretation of quantum mechanics. Subjectivism allows probability assignments to the single case, and represents that mixture of epistemic and objective probability that the founders of quantum mechanics seemed to have in mind.

However, a closer look at de Finetti's philosophy of probability reveals a major reason why his perspective cannot be easily combined with the Copenhagen interpretation of quantum mechanics. De Finetti embraces a positivistic philosophy: "...One could say that my point of view – he says – is analogous to Mach's positivism, where by 'positive fact' each of us means

---

[50] For a discussion of pure and mixed states see van Fraassen 1991, ch. 7.

only his own subjective impressions".[51] Accordingly, probability is for him nothing other than the sensation of an individual.[52] As we have seen, this kind of philosophy was explicitly rebutted by Born and Heisenberg, and their attitude in this connection was certainly shared by physicists of their time. Indeed, crude sensationalism does not even seem to characterize the logical empiricist image of science.[53]

An additional difficulty is related to de Finetti's rejection of the notion of "repeatable experiment" and his claim that one can only talk about single events. Both notions seem to be required within quantum mechanics, for one talks about proper single events, like one particular collision, but one also speaks of single particles in repeated experiments. The possibility of referring probability assignments directly to single events is certainly an advantage of subjectivism over frequentism for interpreting probability in quantum mechanics, but then subjectivism should not banish the notion of repeated experiments.

The reason why de Finetti banished such a notion from his view of probability was philosophical, namely he wanted to build a purely empiricist construction. The will to leave out all metaphysical elements is responsible for the fact that de Finetti did not even want to retain laws and theories. In this connection, he claims that "...one can meaningfully talk only about facts, single facts. Probability can only be assigned to facts, single facts".[54] Laws and theories to him are metaphysical, because it cannot be decided whether they are true or false. They cannot even be assigned a probability, for this would mean making infinite predictions about an infinite number of facts. Instead of engaging in such "absurdities", for de Finetti one should simply give laws and theories a pragmatical import. This amounts to the

---

[51] De Finetti 1931a, p. 171 of the English translation.

[52] See de Finetti 1931b.

[53] Born (1953) charges logical empiricism with sustaining this kind of philosophy: "The logical positivists – he says – who emphatically claim to possess the only rigorous scientific philosophy, as far as I understand, regard the [theoretical] constructs merely as conceptual tools for surveying and ordering the crude sense data which alone have the character of reality" (p. 147). I do not wish to enter into historical details about this issue, but it seems to me that the idea of grounding scientific knowledge on "crude sense data" cannot be taken as something on which all neoempiricist thinkers would have agreed. It was certainly not shared by M. Schlick: see his *Raum und Zeit in der gegenwärtigen Physik*, especially chapter X. Nor does it seem to have characterized the subsequent literature by people like Reichenbach and Carnap, if we leave out the parenthesis of *Der logische Aufbau der Welt* (1928). As a matter of fact, in his "Intellectual Autobiography" Carnap claims that the phenomenistic basis in terms of experience chosen in the book was one among all possible others (see Carnap 1963, pp. 16-19). In any case, he soon turned towards a physicalistic basis in terms of measurements.

[54] De Finetti 1971, p. 89.

fact that a law or theory "induces us to expect that certain facts occur in a way that we take to be in agreement with our idea of that law or theory. The formulation of a theory, of a law, is a link – to a certain extent a deceptive one because it is metaphysical, but often necessary as an attempt towards a synthesis that simplifies complex things – in the mental process through which we pass from the observation of facts in the past to the prediction of facts in the future".[55] In other words, laws and theories can guide our predictions, but we have to bear in mind that they are mere instruments, useful "mental intermediates" between certain facts and certain others. No other role is assigned to theories by de Finetti. In the same fashion, he is not willing to retain any notion of "objective chance", which is dismissed as metaphysical.

Things are quite different with Ramsey. At the beginning of his "Truth and Probability" he says that there are two notions of probability: probability in physics and subjective probability. We also know that at the time of his death in 1930 he was working on a book, also bearing the title *Truth and Probability*, that was planned to include a treatment of probability in physics.[56] Though he did not live long enough to fulfil this plan, one can gather from various notes he left some hints as to his idea of probability in physics. In particular, in his note "Chance" (1928) he not only puts forward a view of "chance" essentially based on degrees of beliefs, but claims that probability in physics means "chance as here explained, with possibly some added complexity ...".[57] Degrees of belief in this case do not represent those of "any actual person" but "a simplified system" including in the first place natural laws "which are ... believed for certain, although, of course, people are not really quite certain of them".[58] Probability in physics is then made to depend of theories. These, in turn, within Ramsey's philosophy are accounted for in pragmatical terms. Clearly enough, Ramsey developed a philosophy that allowed him to combine subjective probability with the idea that within natural science, particularly in physics, probability assignments are somehow determined by theories.

I have argued elsewhere [59] for a possible integration of de Finetti's subjectivism with Ramsey's pragmatist philosophy. Within such a broader framework one could speak of "probability in physics" on a subjectivistic basis. In particular, it would make sense to claim that probability assignments in physics are determined at least partly by theories. Whether this notion of

---

[55] *ibidem.*

[56] More details on this point can be found in my "Introduction" in Ramsey 1991.

[57] Ramsey 1931, p. 96.

[58] *ibidem*, p. 94.

[59] See Galavotti 1991.

probability could be a good candidate for interpreting quantum mechanical probabilities is a matter for further investigation.

REFERENCES

Born, M. (1928), On the Meaning of Physical Theories, in Born (1969), pp. 13-30.

Born, M. (1953), Physical Reality, *Philosophical Quarterly* **3**, pp. 139-149.

Born, M. (1955), Statistical Interpretation of Quantum Mechanics, *Science* **122**, 675-679, reprinted in Born (1969), pp. 89-99.

Born, M. (1965a), Symbol and Reality, *Universitas* **7**, 337- 353, reprinted in Born (1969), pp. 132-146.

Born, M. (1965b), In Memory of Einstein, *Universitas* **8**, 33- 44, reprinted in Born (1969), pp. 155-165.

Born, M. (1969), *Physics in my Generation*, New York, Springer.

Carnap, R. (1928), *Der logische Aufbau der Welt*, Berlin, Weltkreis, English translation *The Logical Structure of the World*, Berkeley, Univ. of California Press, 1969.

Carnap, R. (1963), Intellectual Autobiography, in *The Philosophy of Rudolf Carnap*, ed. by P.A. Schilpp, La Salle, Ill., Open Court, pp. 3-84.

de Finetti, B. (1929), Funzione caratteristica di un fenomeno aleatorio, in *Atti del Congresso Internazionale dei matematici*, Bologna, Zanichelli, pp. 179-190, reprinted in de Finetti (1981), pp. 97-108.

de Finetti, B. (1931a), Probabilismo, *Logos*, 163-219, English translation in *Erkenntnis* **31** ,1989, 169-223.

de Finetti, B. (1931b), Sul significato soggettivo della probabilità, *Fundamenta matematicae* **17**, 298-329, English translation in de Finetti (1993), pp. 291-321.

de Finetti, B. (1931c), Le leggi differenziali e la rinunzia al determinismo, *Rendiconti del Seminario Matematico della R. Università di Roma*, 2nd series,**7**, 63-74, English translation in de Finetti (1993), pp. 323-334.

de Finetti, B. (1936), Statistica e probabilità nella concezione di R. von Mises, *Supplemento statistico ai Nuovi problemi di Politica, Storia ed Economia* **2**, 9-19, English translation in de Finetti (1993), pp. 353-364.

de Finetti, B. (1937), La prévision: ses lois logiques, ses sources subjectives, *Annales de l'Institut Henri Poincaré* **7**, 1-68, English translation in H.E. Kyburg and H.E. Smokler (eds.), *Studies in Subjective Probability*, New York, Wiley, 1964, pp. 95-158.

de Finetti, B. (1941), Punti di vista: Hans Reichenbach, *Statistica* **1**, 125-133, reprinted in de Finetti (1989), pp. 237-248.

de Finetti, B. (1951), Recent Suggestions for the Reconciliation of Theories of Probability, in *Proceedings of the Second Berkeley Symposium on Mathematical Statistics and Probability*, Berkeley, Univ. of California Press, pp. 217-225, reprinted in de Finetti (1993), pp. 375-387.

de Finetti, B. (1971), Probabilità di una teoria e probabilità dei fatti, in *Studi di probabilitá, statistica e ricerca operativa*, ed. by G. Pompilj, Gubbio, Oderisi, pp. 86-101.

de Finetti, B. (1972), Subjective or Objective Probability: is the Dispute Undecidable?, *Symposia Mathematica* 9, 21-36.

de Finetti, B. (1978), Einstein: Originality and Intuition, *Scientia* 72, 123-128.

de Finetti, B. (1981), *Scritti (1926-1930)*, Padova, CEDAM.

de Finetti, B. (1989), *La logica dell'incerto*, ed. by M. Mondadori, Milano, Il Saggiatore.

de Finetti, B. (1991), *Scritti (1931-1936)*, Bologna, Pitagora.

de Finetti, B. (1993), *Induction and Probability*, ed. by P. Monari and D. Cocchi, Bologna, CLUEB (supplement to *Statistica* 52 ,1992, n.3).

van Fraassen, B. (1991), *Quantum Mechanics: An Empiricist View*, Oxford, Oxford Univ. Press.

Galavotti, M.C. (1989), Anti-realism in the Philosophy of Probability: Bruno de Finetti's Subjectivism, *Erkenntnis* 31, 239-261.

Galavotti, M.C. (1991), The Notion of Subjective Probability in the Work of Ramsey and de Finetti, *Theoria* 57, 239-259.

Heisenberg, W. (1925), Über quantentheoretische Umdeutung kinematischer und mechanischer Beziehungen, *Zeitschrift für Physik* 33, 879-893, English translation in van der Waerden, ed. (1967), (all page references to the English edition).

Heisenberg, W. (1927), Über den anschaulichen Inhalt der quantentheoretischen Kinematik und Mechanik, *Zeitschrift für Physik* 43, 172-198, English translation in Wheeler and Zurek, eds. (1983), (all page references to the English edition).

Heisenberg, W. (1930), *The Physical Principles of Quantum Theory*, Univ. of Chicago Press, 2nd edition New York, Dover Publ. Co., 1949 (all page references to the latter edition).

Heisenberg, W. (1955), *Das Naturbild der Heutigen Physik*, Hamburg, Rowohlt, English translation *The Physicist's Conception of Nature*, London, Hutchinson, 1958 (all page references to the latter edition).

Heisenberg, W. (1958), *Physics and Philosophy*, reprinted London, Penguin Books, 1990 (all page references to the latter edition).

Laplace, P.S. (1814a), *Essai philosophique sur les probabilités*, Paris.

Laplace, P.S. (1814b), *Theorie analytique des probabilités*, Paris, 2nd edition.

Margenau, H. and Park, J.L. (1967), Objectivity in Quantum Mechanics, in *Delaware Seminar in the Foundations of Physics*, ed. by M. Bunge, Berlin, Springer, pp. 161-187.

von Mises, R. (1951), *Positivism*, Harvard, Harvard Univ. Press, reprinted New York: Dover, 1968 (all page references to the latter edition).

von Mises, R. (1928), *Wahrscheinlichkeit, Statistik und Wahrheit*, Wien, Springer, English revised edition *Probability, Statistics and Truth*, New York, Dover, 1957 (all page references to the latter edition)

Pais, A. (1982), Max Born's Statistical Interpretation of Quantum Mechanics, *Science* **218**, 1193-1198.

Ramsey, F.P. (1931), *The Foundations of Mathematics and Other Logical Essays*, ed. by R.B. Braithwaite, London, Routledge and Kegan Paul.

Ramsey, F.P. (1991), *Notes on Philosophy, Probability and Mathematics*, ed. by M.C. Galavotti, Naples, Bibliopolis.

Schlick, M. (1917), *Raum und Zeit in der Gegenwärtigen Physik*, Wien, Springer, English translation 1920, reprinted in *Philosophical Papers*, vol. I, ed. by H.L. Mulder and F.B. van de Welde-Schlick, Dordrecht, Reidel, 1979, pp. 207-269.

van der Waerden B.L. ed. (1967), *Sources of Quantum Mechanics*, Amsterdam, North-Holland.

Wheeler, J.A. and Zurek, W.H. eds. (1983), *Quantum Theory and Measurement*, Princeton, Princeton Univ. Press.

Foundations of Science
1 (1995/96), 119-130

Paul Humphreys
Corcoran Department of Philosophy
University of Virginia
Virginia 22903, USA

COMPUTATIONAL EMPIRICISM [1]

Key Words: Computational science, Theories, Models, Scientific instruments, Syntax, Semantics.

Abstract. I argue here for a number of ways that modern computational science requires a change in the way we represent the relationship between theory and applications. It requires a switch away from logical reconstruction of theories in order to take surface mathematical syntax seriously. In addition, syntactically different versions of the "same" theory have important differences for applications, and this shows that the semantic account of theories is inappropriate for some purposes. I also argue against formalist approaches in the philosophy of science and for a greater role for perceptual knowledge rather than propositional knowledge in scientific empiricism.

The widespread use of powerful computational devices in many sciences, often in conjunction with sophisticated instrumentation, is changing scientific method in an essential way. I have adopted the term 'computational empiricism' to describe the general approach underlying these changes. It was chosen as a deliberate parallel to the term 'logical empiricism' so as to emphasize the fact that the logically oriented methods philosophers have used for the greater part of this century need to be replaced, at least for

---

[1] The term 'computational empiricism' was suggested to me in conversation at a philosophy conference in Venice, Italy in June 1991 by someone whose name I have unfortunately forgotten. It seemed to capture perfectly the set of techniques I had described in my talk there, and I have since adopted it. I thank the originator of this term, whoever he is.

some purposes, by methods that are computationally oriented. In fact, as I argue below, this shift in method requires a reversion to the syntactically based representations favoured by the logical empiricists, rather than the structurally based, model theoretic approaches that have gained ascendancy amongst philosophers of science in recent years. However, I want to remain neutral here on the issue of whether these computational methods move us towards an instrumentalist account of scientific representations. I suspect that there is no general answer to this question: some such methods are just predictive and descriptive instruments and nothing more, others can be given a realist interpretation.

I have described some of the changes brought about by the widespread and growing use of computer simulations and related methods in a series of articles (Humphreys, 1991, 1994, (forthcoming) see also Rohrlich, 1991). Here I shall restrict my focus to the most important methodological issues. I want to emphasize that the changes being brought about by vastly enhancing our native computational powers are as important to science as was the earlier development of sensory extenders such as microscopes and infra-red telescopes. These changes are not transitory, as the introduction of new theories tend to be, and they cut across many fields. They affect both theory and experiment, albeit in different ways. I shall lay out my case in the form of five claims, and try to make a plausible case within the limits of this article. (I refer the reader to the articles cited above for further details.)

1. One of the primary features that drives scientific progress is the development of tractable mathematics.

This is such an obvious truth that it would not be worth stating had it not been pushed to the background in methodological studies by the emphasis on metamathematics and logic.[2] This emphasis is in turn the result of the contemporary drive towards abstract representation. There is a natural tendency within the modern axiomatic tradition (or at least that part of the tradition motivated by philosophical interests) to represent theories at the highest level of abstraction possible. This is undoubtedly useful for certain purposes - representing the logic of quantum mechanics by the projection lattice associated with the Hilbert space representation of quantum states is a well-known example [3] - but such an extreme level of abstraction is entirely

---

[2] I am not claiming that this is an inevitable consequence of these logical methods, but it is sufficiently widespread that the emphasis on logical reconstructions within axiomatics must be partly responsible.

[3] See Redhead, 1987, §7.4 for a good account of this tradition; Hooker, 1975 for a collection of (by now classic) papers in the area. Hughes, 1989 is another clear presentation of the Hilbert space approach.

inappropriate for describing the relation between a theory and its applications. One can put the point this way: in von Neumann (1932/1955) it was shown using the abstract Hilbert space representation that the Schrödinger and Heisenberg representations of quantum theory are equivalent. But for applications of quantum mechanics to real systems, various logically equivalent representations are not equally advantageous. For example, when using spherical coordinates θ, ϕ for the treatment of the hydrogen atom, there are no analytic functions of θ and ϕ for the Schrödinger representation that give half-integral values for the angular momentum. In contrast, a pure matrix treatment of angular momentum provides the correct representation for the states of both integral and half-integral values.[4] Alternatively the interaction representation, due to Dirac, is better suited to S-matrix computations for time dependent approaches to collision phenomena than is the Schrödinger representation. However, in special cases, such as field theories with a nonlocal interaction, the Heisenberg representation has to be used.[5] But of course, even this level of generality is far too abstract to provide us with concrete solutions in many cases. To employ Schrödinger's equation ($H\psi = E\psi$) some specific Hamiltonian must be used for the system under investigation, and selection of even a moderately realistic Hamiltonian for relatively simple atoms or molecules quickly takes us outside the realm of exact solutions. The same point holds for using the general form of Newton's Second Law ($F = ma$), where the choice of a specific force function F usually leads us outside the realm of analytically solvable equations.[6] The point is not that it is the complexity of the world which leads to the unsolvability of its mathematical representations, although complexity does often produce this. It is that even if the world were quite simple, in the sense that certain sorts of simple equations [7] represented its structure exactly; we should still be forced to adopt approximate or numerical solutions to our (true) theories of the simple world. This is why the widespread use of computational methods in various sciences, which now form an indispensable set of core approaches in physics, chemistry, molecular biology and, to a lesser extent, in the social sciences, is at least as important an advance as was the invention of the differential and integral calculus. These methods expand enormously the number and types of theories that can be brought

---

[4] See e.g. White, 1966, pp. 193-201.

[5] See Roman, 1965, §4.3.

[6] For specific examples in these two cases, see Humphreys, 1991, pp. 498-99.

[7] Such as the Lotka-Volterra equations in population biology ($dx/dt = ax + bxy, dy/dt = cy + dxy$) or a second order, non-linear homogeneous differential equation describing the fall of a body under gravitation and the action of a drag force dependent on the square of the body's velocity.

to bear in a practical way on specific applications. What is important here is not computability in principle, but computability in practice, and for the purpose of testing and confirming theories, it is only the latter that matters. In this regard Laplace's famous epithet about Laplacian mechanics has its ironic side.[8] Laplace's omniscient entity is given an initial value problem to solve. But Laplace's equation with open Dirichlet or Neumann boundary conditions has no unique solution unless special conditions at infinity are assumed, and so Laplace's calculator might well not be able to solve Laplace's equation exactly unless it was endowed with access to analytic methods of mathematics to which we do not have access.[9]

There are many examples of the development of new mathematical techniques extending the domain of application of scientific theories ranging from the very general, such as the introduction of the differential calculus and the development of contemporary statistical theory, through the more localized techniques such as spherical trigonometry and coordinate geometry, to the mundane, but important, invention of logarithms. Without such apparatus available, contemporary science in many areas would be vastly cruder and less developed than it now is.

We can draw a second moral from the importance that specific syntactic representations have for applications:

2. The semantic account of theories cannot represent what is important for the application of a scientific theory to real systems.

Recall that one of the primary advantages of the semantic account (see e.g. Suppes (1970), van Fraassen (1972)) was that by virtue of focussing on the class of models that represents the abstract structure of the theory, rather than on the particular accidental syntactic forms that are used to linguistically display that theory, inessential linguistic differences are given no weight. To take a simple example, one can axiomatize probability theory syntactically by the axioms of non-negativity, normalization, and countable additivity. If, instead, one replaces the countable additivity axiom with the axiom of finite additivity and the axiom of continuity $(E_n \downarrow 0 \Rightarrow P(E_n) \to 0)$, the same class of models is arrived at. But this devaluation of the particular syntactic form used is wrong if one is interested in how theories are applied and empirically tested. As we have seen, to bring a particular theoretical

---

[8] "Given for one instant an intelligence which could comprehend all the forces by which nature is animated and the respective situation of the beings who compose it - an intelligence sufficiently vast to submit these data to analysis - it would embrace in the same formula the movements of the greatest bodies of the universe and those of the highest atom; ... (Laplace, 1795)

[9] See Morse and Feshbach, 1953, 6.1, 6.2; also Guenther and Lee, 1988, §8.5.

representation into explicit contact with data often requires transformations from one syntactic version to another in order to move from unsolvable to solvable treatments. As heavily computational approaches to theory applications become increasingly important, so will the need to give a proper place to theories as pieces of syntax that are not equivalent viz-a-viz applications.[10]

Our third claim is this:

3. Mathematically formulated theories usually come with an interpretation attached that justifies their employment, and any account of theories that splits the formalism from its interpretation misrepresents this feature.

Formal treatments of scientific theories frequently treat them (viewed now as pieces of mathematical syntax) as uninterpreted calculi from which any original interpretation can be peeled off, and another consistent interpretation imposed on the bare formal bones. If one does adopt such a purely formal attitude towards theories (an attitude which Hilbert, for example, rejected in such an extreme form for mathematics itself) then one is immediately faced with a version of the *underdetermination problem*: there are always multiple interpretations of the formal theory available each of which is a model for that theory i.e. each of which makes the theory "true" by virtue of satisfying its axioms. Thus, the same formal theory can be given a variety of distinct interpretations, all of which are compatible with the empirical data and the formal constraints imposed by the theory, but each of these interpretations employs a different ontology.

It is this idea that one can strip off whatever interpretation originally came with a scientific theory and reinterpret it by means of some structurally identical, but ontologically distinct, model that provides one of the most powerful arguments for underdetermination in science and concomitantly an argument against realism. This sort of formalism is a caricature of scientific theories, for, with occasional exceptions, the relationship between the fundamental principles of a science and the specific applications which make the theory testable, and hence scientific, requires a specific interpretation in order for the specific application to be derived. In consequence

---

[10]I note for historical accuracy here that Suppes does allow a practical role for syntax (see Suppes, 1970, p. 2-18ff) but asserts that even for empirical tests of a theory, a standard (first-order logic) formulation is not necessary. Unfortunately, the chapter within which this is explicitly discussed is not in the circulated manuscript. It is worth mentioning here that the fascination with first order logic evinced by many philosophers results in set of mathematical apparatus that is far too restricted for the needs of theoretical science. Even such elementary concepts as 'random variable' and 'having a probability greater than some real number p' cannot be adequately represented in first order logic. (See Barwise and Feferman, 1985, Chapter 1.)

the 'intended interpretation' of a theory cannot be lifted from that theory without the justification for the derivation of its empirical consequences disappearing.

This is easiest to see in the case of theories couched in the form of differential equations.[11] Such theoretical representations, in the form of diffusion equations, laminar flow equations, and so on, although frequently used as "off the shelf models", are generally not mere instrumental devices justified in terms of their utility,[12] but are constructed from fundamental principles, ontological assumptions about the constituents of the system and their properties, assumptions about boundary conditions, and various approximations and idealizations. At least the last three of these cannot be justified without the use of some specific interpretation of what kind of system is being represented. The kinds of approximations that are routinely used by physicists to justify the use of a mathematical limit, and which grate against mathematicians' finer sensibilities, would be completely ad hoc were it not for the concrete interpretation imposed during the derivation. Considerations such as the number of particles in the system, whether a conductor is perfectly insulated, order of magnitude approximations, the finiteness of ratios as the numerator and denominator go to zero, smoothness assumptions, and so on, are all integral to the derivation of such equations.

To reinforce this point, consider that all theories that employ such approximations and idealizations will not fit the data exactly. A primary reason for not rejecting those theories is precisely the physical justifications used in constructing the theory which show that such deviations from the data are to be expected. Removing the interpretation from the theory would take away the justification for retaining the theory in the face of falsifying data.

These considerations take on a greater prominence in the back-and-forth fitting of models to data in computer simulations. The almost instantaneous adjustment of boundary conditions, parameter values, theoretical assumptions, time-step size and so on that is used to bring the simulation into conformity with the data would be completely unmotivated without the intended interpretation being used. I want to emphasize that this role of interpreted theories is not peculiar to computationally implemented theo-

---

[11]It is also easy to establish in the case of statistical models, although I shall not do that here.

[12]Some are, of course. Realism is not a blanket interpretation that one imposes on all scientific theories in a uniform way. Some theories are meant to be interpreted in a realistic way, others are explicitly intended only as instrumental devices, and it is usually quite clear which are which. The same point applies to terms within theories; many parameters are meant to have a realistic interpretation, others not. Realism is a weak thesis, though. One needs only a few (or even one) realistically interpreted theories to be true for (local) realism to hold.

ries however. It is a perfectly general feature of model construction and it has the crucial consequence that it renders deeply suspect the kinds of reinterpretation strategies that Carnap and his descendants, such as Quine, Goodman, and Putnam have used to attack realism. (As in, for example, Carnap (1956); Quine (1969), Goodman (1978), Putnam (1980).)

Let us now consider some other, though no less important aspects of computational methods. We can describe one kind of computational science that clearly constitutes a genuinely new kind of method. Suppose that we are faced with the task of computing values of the equilibrium distribution for energy states of a canonical ensemble of particles, which will be represented by the Boltzmann distribution $\exp\left[-H(x)/K_\beta T\right]$. Frequently the values of this will not be known for a given Hamiltonian H, and Monte Carlo methods need to be employed to calculate them. One common method, using the Metropolis algorithm, constructs a random walk in the configuration space of the ensemble of particles, the random walk having a limit distribution that is identical to the one required. The important feature of these random walks is that they are run on real computational devices within numerical models of configuration space. Although the process is carried out numerically, the stochastic process is based on a specific physical model of the system. But the limit values are gained neither by traditional empirical experimentation on the real system that has been represented, nor by analytic methods of traditional mathematics. Thus, the only way to calculate the unknown limit distribution is to run repeated trials of the numerical stochastic process and let the computational device dynamically produce the limit values. These methods thus occupy an intermediate position between physical experimentation and numerical methods.[13]

A comparison with Buffon's needle process is illuminating.[14] Although this process is perhaps best known to philosophers for having induced Bertrand to invent his paradoxes of equiprobability, Buffon's original process is often considered to be the progenitor of empirical Monte Carlo methods. In its simplest form, the process consists of casting a needle of length L onto a set of parallel lines unit distance apart. (Here $L \leq 1$). Then, by an elementary calculation, one can show that the probability that the needle will intersect one of the lines is equal to $2L/\pi$. As Laplace pointed out, frequency estimates of this probability produced by throwing a real needle can be used to empirically estimate the value of π, although as Gridgeman (1960) notes, this is a hopelessly inefficient method for estimating π, since if the needle is

---

[13]In fact, it has been claimed that the Monte Carlo method is useful only in these cases, otherwise we should simply be checking the random number generators.

[14]This paragraph is taken verbatim from Humphreys, 1994.

thrown once a second day and night for three years, the resulting frequency will estimate π with 95% confidence to only three decimal places. Be this as it may, the Buffon/Laplace method is clearly empirical in that actual frequencies from real experiments are used in the Monte Carlo estimator, whereas the standard methods for abstractly approximating the value of π using truncated trigonometric series are obviously purely mathematical in form. In contrast, the dynamic implementation of the Metropolis algorithm lies in between these two traditional methods. It is in this sense at least that I believe numerical experimentation warrants the label of a new kind of scientific method. Thus we have:

4. Some methods of computational science constitute a genuinely new scientific method, intermediate between traditional experiment and abstract theoretical calculation.

As the last claim we have:

5. Extending our native mathematical abilities by means of computational devices is analogous to extending our natural perceptual abilities by scientific instruments.

Computer simulations can be compared to a microscope of variable power. By decreasing the grid size (the time step or the spatial increment) in finite difference methods, or by increasing the sample size in Monte Carlo methods, increased detail about the solution can be obtained. The degree of detail available is often limited by technological constraints such as the speed of the computational device and its memory capacity. Limitations of this kind often mean that three dimensional systems have to be simulated in two dimensions, or that the dimensionality of the phase space used is significantly reduced relative to the dimensionality of the real system's state space.

We can pursue the analogy a little further. In the case of microscopes, Ian Hacking (Hacking (1983) Chapter 11) has noted that it is often possible to directly compare the object being observed with the result of using the instrument on that object, in order to check the veridicality of the instrument. In the example that Hacking uses to argue for this point, a photographically reduced grid is examined under the microscope and the result compared with the original. Although one needs to be careful with this argument, because it requires that the instrument in question can inspect the same aspect of a (reduced) system as can the human eye, and that a given microscope can be used on a system amenable to such reduction techniques (if the microscope could not be used to inspect objects larger than a few atoms across, Hacking's technique could not be used directly), it can be supplemented with the same kind of cross-checking transitivity argument that one uses to avoid extreme operationalism. Why do different kinds of thermometers measure the same quantity? Because in temperature regions where they overlap, they always, within measurement error, give the same result. By moving up the scale of temperature across successive regions of overlap, one can argue that spectral analysis thermometers measure the same quantity as do alcohol thermometers. Similarly with microscopes, from ordinary optical bench models down to scanning tunneling devices. And so with simulations. It is the cross-checking against analytically derived solutions and empirical data in validation studies that is crucial to the acceptance of a particular simulation. This is one reason why, where it is available, numerical analysis is so important. Having a theoretical justification for convergence and stability results is a powerful justification for the validity of a given numerical method. That is why, with many combinatorially complex calculations for which we have no cognitive access to the result independent of the simulation itself, comparison with already existing, verified techniques is crucial, just as justification of the use of other instruments takes place not in isolation, but in comparison with the overlap with measurements from established devices.

One other comparison between microscopes and simulations is useful, and hinges on the ability of the latter to process quantities of (empirical) data that are far too large, taken item by item, for humans to cognitively assimilate. This function is enhanced by the ability of some modern measuring devices to generate measurements much faster than human observers could tabulate them. When one thinks of traditional optical devices such as optical microscopes and telescopes, the tendency is to think of the data being presented directly to the observer, but many optical systems now use enhancement techniques to produce their images; indeed, many 'photographs' in mass circulation magazines are enhanced in this way to improve definition. More directly relevant is the fact that many types of microscopes and other instruments are now combined with computational algorithms to produce their images. The output from a scanning electron microscope can be digitized, and various global or local features such as the volume fraction and grain sizes of alloys, and the brightness and image texture contrasts of specific features can be exhibited.[15] This coupling of computation and instrumentation renders suspect Hacking's claim [16] that it is not necessary to know how a microscope works in order to use it. It is precisely because specific information has to be extracted from a given image that the user has to know which algorithm is appropriate for analyzing which feature. This

---

[15]See e.g. Williams *et al*, 1991, Chapter 12.

[16]Hacking, 1983, p. 191.

information extraction problem is in principle no different from the better known case of interpreting visual images using background knowledge to pick out specific features.

It should not be thought that simulation methods are unique in their need to rely on discrete approximations of more finely-grained phenomena, these approximations being quite coarse with respect to the real level of detail involved. For example, palaeontologists can ordinarily differentiate fossil records at approximately 100,000 year intervals. Anything on the evolutionary scale that occurs on a smaller time interval than that is undetectable. This, of course, potentially means the loss of enormous detail in the evolutionary sequence (recall how many species have become extinct in the last 100 years alone) yet one can still draw legitimate large-scale structural inferences based on the available level of differentiation.

Of course, this analogy takes on a literal construal when the results of computer simulations are run on graphics displays. Then the level of increased detail can literally be seen. One important feature of simulation testing is to preserve the principal structural features of the system as the level of detail is increased, thus confirming that those features are stable solutions, rather than artifacts of the approximations used.

It is worth remarking that with the easy transition between propositional representations and graphical representations that is a characteristic of computational methods, the traditional emphasis upon propositional representations of the comparison between theory and data is seen to be unnecessary. This is not just a pragmatic fact that far more information can be assimilated by humans when it is presented graphically than in a propositional representation, but an epistemological claim: that a direct comparison can be made between the output of the simulation and the structure of the system being modelled. This suggests that a more central role be given to direct perceptual confirmation of theories within these contexts, rather than the heavy orientation towards propositional attitudes, such as beliefs. In many cases, the output from a simulation can simply be *seen* to fit the data.[17]

References

Barwise, J. and Feferman, S. (1985), *Model-Theoretic Logics.* New York: Springer-Verlag.

Carnap, R. (1956), Empiricism, semantics, and ontology, pp. 205-21 in his *Meaning and Necessity* (2nd Ed.). Chicago: University of Chicago Press.

[17] A preliminary version of this paper was read at the first Association for the Foundations of Science Workshop, Madralin, Poland in August 1994. I thank the participants of that splendidly cooperative venture for many helpful comments and suggestions.

Goodman, N. (1978), *Ways of Worldmaking*. Indianapolis: Hackett Publishing Company.

Gridgeman, N. (1960), Geometric Probability and the Number π, *Scripta Mathematica* **25**, 183-195.

Guenther, R. and Lee, J. (1988), *Partial Differential Equations of Mathematical Physics and Integral Equations*. Englewood Cliffs: Prentice-Hall.

Hacking, I. (1983), *Representing and Intervening*. Cambridge: Cambridge University Press.

Hooker, C.A. (ed.) (1975), *The Logico-Algebraic Approach to Quantum Mechanics Volume I: Historical Evolution*. Dordrecht: D. Reidel Publishing Company.

Hughes, R. I. G. (1989), *The Structure and Interpretation of Quantum Mechanics*. Cambridge: Harvard University Press.

Humphreys, P. (1991), Computer Simulations, pp. 497-506 in *PSA 1990*, Volume II, A. Fine, M. Forbes, and L. Wessels (eds). East Lansing: Philosophy of Science Association. – (1994), Numerical Experimentation, pp. 103-121 in *Patrick Suppes: Scientific Philosopher*, Volume 2, P. Humphreys (ed). Dordrecht: Kluwer Academic Publishers. – (forthcoming), Computational Science and Scientific Method in *Minds and Machines*.

Laplace, Pierre Simon Marquis de (1795), *Essai philosophique sur les probabilités*. Translated as *A Philosophical Essay on Probabilities*. New York: Dove Publications (1951).

Morse, P. and Feshbach, H. (1953), *Methods of Theoretical Physics*, Part 1. New York: McGraw-Hill, Inc.

Putnam, H. (1980), Models and Reality, *Journal of Symbolic Logic* **45**, 464-82.

Quine, W. V. O. (1969), Ontological Relativity in his *Ontological Relativity and Other Essays*. New York: Columbia University Press.

Redhead, M. L. G. (1987), *Incompleteness, Nonlocality, and Realism: A Prolegomenon to the Philosophy of Quantum Mechanics*. Oxford: The Clarendon Press.

Rohrlich, F. (1991), Computer Simulations in the Physical Sciences, pp. 507-518 in *PSA 1990*, Volume 2, A. Fine, M. Forbes, and L. Wessels (eds). E. Lansing: Philosophy of Science Association.

Roman, P. (1965), *Advanced Quantum Theory*. Reading (Mass.): Addison Wesley Publishing Company.

Suppes, P. (1970), *Set-Theoretical Structures in Science*. Mimeo manuscript, Institute for Mathematical Studies in the Social Sciences, Stanford University.

van Fraassen, B. (1972), A Formal Approach to the Philosophy of Science, pp. 303-366 in *Paradigms and Paradoxes*, R. Colodny (ed.). Pittsburgh: University of Pittsburgh Press.

von Neumann, J. (1932/1955), *Mathematische Grundlagen der Quanten- mechanik*. Berlin: Springer. Translated as *Mathematical Foundations of Quantum Mechanics*. Princeton: Princeton University Press.

White, R.L. (1966), *Basic Quantum Mechanics*, New York: McGraw-Hill Book Company.

Williams, D., Pelton, A., and Gronsky, R. (1991), *Images of Materials*. Oxford: Oxford University Press.

Foundations of Science
1 (1995/96), 131-154

Joseph Agassi
Department of Philosophy
York University, Toronto
Ontario M3J 1P3, Canada

BLAME NOT THE LAWS OF NATURE

> A theory is lucky if it gets some of the
> results right some of the time.
>
> Nancy Cartwright[1]

Abstract.
1. Lies, Error and Confusion
2. Lies
3. The Demarcation of Science: Historical
4. The Demarcation of Science: Recent
5. Observed Regularities and Laws of Nature

1. What the public-relations spokespeople say is regularly in the gray area between ignorance, confusion and lies.

2. The integrity of science rests on precarious foundations. The traditional identification of science with proof, without any theory of proof to back it up, today simply harbors danger.

3. Only two rules, Boyle's and Newton's, are generally admitted as inherent to science. They functioned remarkably well, but present conditions require new measures.

4. Popper's demarcation of science is partial: refutability is empirical character, and not every empirical research is scientific.

---

[1] *How The Laws of Physics Lie*, p. 165.

5. We do not know what in the life sustaining conditions is due to the laws of nature and what is due to mere local regularities.

1. Lies, Error and Confusion

Book titles may be very influential. The title of recent book by Nancy Cartwright is by far better known than its contents; it is, *"How The Laws of Physics Lie"*. Now Professor Cartwright knows full well that the laws of physics can do nothing except, in a sense, regulate or rather constrain the conduct of things natural; they prevent the occurrence of miracles proper (which is why they are considered supernatural). Indeed, Professor Cartwright speaks not of laws, of course, but of statements. Statements, too, cannot lie, she knows equally well. Only people can lie. And for this they have to assert some statements, to transmit assertions to some people to whom they may lie. Moreover, in order to lie one has to assert to them at least one statement which (rightly or not) one presumes to be false while pretending to presume that it is true. Finally, this pretense should be intentionally concealed from the audience to whom one lies, as is evidenced in the theater, where the assertion need not be a lie just because the pretense is shared with one's audience.

The statements of the laws of physics, being statements, have no intent. Moreover, by definition, (the statements of) the laws of physics are true. What we think are (statements of) the laws of nature we also think are true – even though, we know full well, they may very well be false: we know full well that we may be in error. But error is no lie. Thus, when the assertion was repeatedly made that Newton's theory of gravity is a law of nature, and so it was also repeatedly suggested that it is true, this was an error, of course, but no lie, of course.[2]

Strangely or not, this last paragraph, evident as it seems, raises some serious problems for many scientists and philosophers, though their number is (hopefully) on the decrease. They find it hard to admit of the great Newton that he was in error. Now if they are not foolish they will easily concede that Newton made many mistakes in his life. If they are even somewhat familiar with the history of science, then they know that he believed in alchemy and in the prophecies of the book of Daniel, and so on. At least they should know that he honestly admitted absolute space and time, both denied by

---

[2]For all this see Agassi, J. (1990/91). For a detailed, recent history see Agassi, J. (1990a).

Einstein. Nevertheless they find it hard to ascribe tc him a false scientific theory, because they cannot allow that science and error can mix. Indeed, though they acknowledge the great difference between falsehood and lies in general, they do not acknowledge it in the context of science. This makes sense: if one merely thinks x, or merely conjectures x, and then claims to be in possession of knowledge of x, then it may very well be the case that one is lying. As science is knowledge (the words "science" and "knowledge" are etymologically synonymous), to proclaim that a statement is scientific is to foster a very heavy burden.

This is true: the assertion that x is far from being synonymous with the assertion that one knows that x, as it is synonymous with the assertion that one conjectures that x: surely a conjecture differs from a knowledge claim. Nevertheless, there is a historical fact here that should not be overlooked: Newton was convinced that he knew the statement known as the inverse square law, and his knowledge claim means that he took the inverse square law to be proven, so that it will never be refuted. And it was refuted in the sense in which he said it will not be. Hence he was in error. Hence, even though in ever so many cases one lies when saying that one knows that x instead of saying that x, this is not true of Newton: he did not lie.

This is not peculiar to science or to its history or its philosophy. In civilized society it holds for witnesses in law-courts regularly. There are diverse legal standards of evidence, and they include diverse standards of the legitimate use of the verb "know", diverse standard rules of their proper employ. Thus, when a journalist wishes to publish some story that may cause damage, they need corroborating evidence. Even if they have sufficient evidence to render their action legitimate, in civilized society the police are not always allowed to act on it: the standard of evidence required before the police can act are usually more stringent than those required of the journalist. For the district attorney to act on police evidence, compliance with even more stringent standards is required. The most stringent of all standards is placed on the jury convicting a party of capital punishment, especially if this leads to the death sentence. Yet this is not to say that juries do not err. Nor is it to say that the standards by which they are to judge are presumed to be well understood and well worded. Proof: these standards are repeatedly revised.[3]

Until recently most philosophers and scientists (Newton included) assumed that the standards of proof within science are the highest possible.

---

[3]For all this see Agassi, J. (1985a). See also Agassi, J. (1987).

All that they said of this proof is that it is inductive. What this word exactly means they did not say, except that it rests on experience. On this they were in error, all of them: there is nɔ inductive proof, only inductive disproof. As Newton thought he knew that the inverse square law is true, he was in error; yet of course he did not lie in the least.

Cartwright is in full agreement here. She says, "A theory is lucky if it gets some of the results right some of the time." This holds for any theory, hence also for a scientific theory: no theory is exempt from possible error. This last claim is called "fallibilism". Hardly any philosopher of science today denies fallibilism (though many are cagey about this fact). But fallibilism is a problem rather than a solution: if there is no proof in science, what distinguishes scientific conjecture from any other? In other words, what are the standards that permit one to say of a conjecture that to is a scientific theory? (Often but not always the word "theory" is distinguished from the word "conjecture" to mean "scientific conjecture".) This is the traditional problem of demarcation of science, put in the new, fallibilist manner: which conjecture is scientific? We will return to this question later, as it is difficult and requires some preparation, some background material. Here we merely glanced at the much simpler matter, the fact that the old standards do not ever apply. This ensures the reasonableness of the view that the inverse square law is false so that it can be asserted without giving offense.

Nevertheless, things are not so simple, and so some modern thinkers who know about the refutations of the inverse square law, and who admit them as true, nevertheless also declare that statement true. They thus seem clearly inconsistent. Are they? It is hard to judge. Perhaps they use the word "true" in a non-standard way, such as the way we use it when we say that it is true of Samuel Pickwick that he was a bachelor but false that he was married. Since Mr. Pickwick is a product of Charles Dickens' imagination, it is impossible both to consider true the statement that he was a bachelor and to consider false the statement that he was married – except in some non-standard sense of the word. One sense in which it is quite proper to consider both the one statement true and the other statement false is rather easily available: it is the sense known as fictional, or *façon de parler*: we talk *as if* the story which Dickens has told is true. (It is not *literally* true, yet we talk *as if* it were.) What is kosher for Dickens surely may be kosher for Newton: what is true or false of Mr. Pickwick may also be true or false of the inverse square law. One may admit then the refutations of that statement yet insists that it is true in the same moderately extraordinary and obviously non-standard sense of the word in which it is true that Mr. Pickwick is a bachelor and not that he is married, in the sense known as fictionalist or "as if" or *façon de parler*.

The idea that a scientific theory is not to be taken literally raises a serious question. What is the point of it? Just to save a theory from refutation is not sufficient: its honor may be saved, but it then will not serve as a theory of the world, and what else is it good for? Now it is traditional wisdom that a theory is both a description of the world and a useful instrument for prediction. The claim that a theory is not to be taken literally, that it has no literal meaning at all, leaves it as an instrument for prediction. The idea that a scientific theory is *nothing but* an instrument for prediction is called *instrumentalism*. Instrumentalists may endorse different theories even if they are inconsistent with each other: denuding them of their informative contents eliminates the contradiction between them. Thus, Newton's theory of gravity contradicts Einstein's, but denuded of their informative contents they are simply different sets of differential equations. The only question regarding them is, which of them is it advisable to use in which conditions? This question is answered in a pragmatic manner.

What, however, should be the verdict on those who admit the refutations of the inverse square law, and admit that they contradict that statement, yet they insists that it is true in the ordinary, standard sense of the word? The German philosopher Georg Friedrich Wilehlm Hegel, as well as some of his followers, including perhaps Karl Marx and Friedrich Engels, are repeatedly reported to have behaved in this manner. It is hard to decide if they really did, since they said that contradictions are true without adding that by "true" they meant "true in the usual sense of the word" (nor that they meant it as "true in an unusual sense of the word"). But they offered examples; and these are most unusual: any changing thing, they said, is self-contradictory, so that a fruit, for example, contradicts the flower from which it has emerged. It is clear that the word "contradict" here is not used in the standard manner, no matter in which manner exactly it is used (if any). Marx said that capitalism suffers from incurable inner contradictions. Whether incurable or not, he clearly meant inner conflicts: the short-term interest of entrepreneurs is to compete as hard and fast as possible, yet this leads to cycles of increasingly severe economic recessions that are not in their interest at all. This makes excellent sense (else it could not be empirically refuted as it was). It is no more a matter of contradictions than that what the laws of nature tell us is a lie: the words "contradiction" and "lie" are used here merely metaphorically.[4]

Still, of those who insist point-blank that they intend to contradict themselves usually cause no trouble, as long as it is asserted that what they say
then is false, since, demonstrably, every contradiction is false. Those who
contradict themselves, then, are plainly mistaken and there is nothing more
to their insistence than to any insistence on any error. Perhaps the error is
due to ignorance, perhaps to confusion, and perhaps it is asserted as a lie;
perhaps those who assert it display a bit of each of these options. Consider
the case of the self-appointed public-relations spokespeople of science (who
are often honest admirers of science, often failed scientists or former scientists with a rich research career behind them). They speak regularly about
these matters, and, as public- relations spokespeople they try to cultivate
public relations, and then what they say is regularly in the gray area between
ignorance, confusion and lies.[5]

2. Lies

Is no one here simply lying? Are some of the public-relations spokespeople of science not simply liars? Hardly. Scientists proper (usually) do
not lie about matters scientific (not even the liars among them). (Public
relations spokespeople of science often pretend that the moral standards of
scientists are generally very high; since most scientists are academics, this
is obviously false: everyone even vaguely familiar with academic life takes
for granted that intrigue is rampant there; all "realist" novels and plays and
movies about academic life present the academic as a stereotype cowardly,
gray male who regular engages in intrigue and thus in deception.)[6] Science
and lies do not mix. This is taken for granted by most philosophers. Taken
literally this is obviously false, as is attested by the information admitted
in the science literature, namely that this literature repeatedly includes lies.
Not all of the lies included in that literature are admitted there to be lies:
some of the lies published there are treated quietly, and then quietly con-
signed to oblivion; only some of them are exposed – often for the first time
in law-courts and then in that literature. How then does that literature
nonetheless deny that science includes lies? It does so on the supposition
that this phenomenon is transitory and marginal, one that can be eliminated

---

[4]See the classical "What is Dialectics?" by Karl Popper, reprinted in his *Conjectures and Refutations*.

[5]For the public relations of science see Agassi, J. (1995a) and Agassi, J. (1995b)

[6]For the academic lifestyle see Agassi, J. (1986a) This is not the place to discuss the merits of the academy despite its defects here depicted. For that see Agassi, J. (1991a).

easily and rapidly. Indeed, the phenomenon is taken to be transitory and marginal on the strength of the supposition that it can be easily and rapidly eliminated. What can be easily eliminated usually is. These suppositions are thus far not refuted. Are these suppositions always true? Will this always be the case? If not, should we not examine the conditions under which this is the case and safeguard these conditions?[7]

There was one period when the scientific community was had: the Atomic Energy Commission of the United States misled all those who opposed to the US government nuclear policy by intentionally misleading the general public, including most nuclear physicists, and this could not be achieved without the cooperation of some nuclear physicists. Moreover, at no time did the leadership of the scientific community as such openly and clearly protest and declared the matter in need of public inquiry. Some scientists and some bureaucrats did clearly protest, and their act demanded great courage. The deception in question was large scale, and some of its repercussions are still matters of public debate and policy, which will not disappear from the agenda for quite a while as yet. Perhaps it was this very deception and the absence of a sharp official protest against it that has led to the deterioration of scientific ethics of the recent decades. Nevertheless, somehow this whole matter is taken to be a passing episode, relating to past issues that have caused damage that is now more-or-less under control.

Just when it was hoped that things are getting under control, another incident took place. The officials in charge of the recent, lamentable star-wars project lied bluntly about their first experiment and they were caught red-handed and publicly exposed. Nothing much happened then. Science went its own way and left the government of the United States to sort things out for itself. Fortunately, the relaxation of global tensions came about anyway, despite star wars (even though that miserable project helped push Russia into chaos a bit more than was unavoidable were the West a bit more flexible). One way or another, all this was due to new developments, to developments resulting from the devastation of Hiroshima by nuclear weapons. It was still hoped that outside the sphere of science-based military technology things are still different, that military technology still is marginal to science even though it is science based. Is that the case? Is science generally more reasonable and honest than governments? If yes, was it always so? If

---

[7]On the matter of imposture in the scientific literature see Laor, N. (1985). See also Agassi, J. (1985a), op. cit.

not, how did all this integrity come about? How is it maintained?[8]

The integrity of science is scarcely maintained. It often rests on precarious foundations. The fact that science was traditionally identified with proof, without any theory of proof to back up this claim, is an obvious fact. (Proof theory proper was possibly invented by David Hilbert in the early twentieth century; more likely it was invented by Kurt Godel in the 'thirties.) Moreover, theories were regularly refuted left and right, yet the paradigm was Newton's inverse square law that survived attempted criticism for two centuries; refuted theories were dismissed off had, or surreptitiously modified, or declared mere instruments for prediction. This is clearly a biased conduct and it is still current.[9] Nay, it is more rampant today, when science is a big business, than ever before. This is not exactly trustworthiness. And yet, somehow, we all know, science has exhibited, all the way, a remarkable degree of integrity and trustworthiness. How? What kind?[10]

Before discussing the great integrity of science we further should examine the question, how virtuous is it really, how trustworthy? Moreover, the question is put a bit too bluntly. It is obvious that science comprises of many complex institutions and traditions and subjects, and it is a bit too careless to assume tacitly, and to take on faith, that the different aspects of it exhibit the same degree of honesty, especially after it was noted that possibly science-based military technology is more military than scientific. Yet integrity is taken as the most obvious characteristic of all science, and even of the university as a whole, even though not all academic activity is scientific and even though not all academic departments claim to be scientific. (Only in the German speaking university every department claims to be scientific: in German the expression *dogmatic science* does not sound as ludicrous as it rightly sounds in English.) Moreover, the university as it was traditionally known for centuries was never famous for integrity. The very canons of scientific integrity came about only in the scientific revolution of the second half of the seventeenth century. They were implemented not in the university system, nor even in any single university; no university has ever endorsed them officially. They were implemented in the Royal Society of London and were then endorsed by all the other modern learned societies, and this was a success that led to the development of the scientific ethos to the point that the universities had to make compromises with the scientific societies and

---

[8]See Agassi, J. (1993a).
[9]See Agassi, J. (1963).
[10]See Agassi, J. (1988a).

recognize them *de facto*. And the university system endorsed the rules of scientific etiquette of the scientific societies only after the secularization that took place after the American and French revolutions, and more clearly so only after Hiroshima. But to this day and age this is not universally true.

A recent instance should illustrate this. When Cardinal Hans Kung was demoted by the current Pope he was appointed professor at a respected German university, in a theology department that is Roman Catholic. His peers objected to the appointment: the new professor is a dissenter and his dissent creates discord and discord destroys unanimity and this threatens the scientific status of the whole field.

It is amazing how narrow academic politics can be: the authority of a theology department rests on a unanimity that is maintained by exclusion as well as by the pretense that there are no other theology departments where different doctrines are taught! Lest one thinks that this is unique to theology, let me report that it is common knowledge in physics circles that impediments are piled on the way of dissenters who wish to read papers in physics conferences, and when all else fails, they are given the opportunity to speak very early or very late in some obscure seminar room. But this is perhaps less characteristic of science proper than of theology. What, however, is science proper? This is the problem of demarcation of science.

Before we come to the demarcation of science we should conclude with the matter of integrity, or rather its absence, or rather the way academics behave when they wish to perpetrate lies. Academics need not lie to be disingenuous; they may simply apply different standards to different cases; they may simply pretend that certain events do not occur. If the establish ment treats ideas one way if they are advanced by its members and another way if they are advanced by outsiders, then its conduct is already in the wrong, and scholarship may suffer from this conduct. According to a few witnesses, an idea advanced by outsiders take two decades or so to reach the recognition the deserve. This is obviously an insufficient evidence. It is no doubt true, as Thomas Kuhn observes, that in physics the time lag is much shorter than in other fields. Yet Kuhn, too, beautifies: when a criticism of an ongoing project is advanced, and the livelihoods of scores of researchers depend on it, there is bound to be tremendous hostility to the critic. In the research related to the pharmaceutic industry one should particularly expect this to be the case, and there is ample evidence that indeed it is, but the establishment prefers to treat infringements piecemeal rather than attack the problem frontally, with the best research tools available.[11]

In the philosophy of science the time lag is, indeed, decades. The two greatest philosophers of science in the mid-century, Michael Polanyi and Karl Popper were ignored by the establishment, because it was still engaged in its advocacy of its standards of infallibility in science, or at least as near infallibility as possible, instead of admitting that this was a bad job. And then Kuhn came with an establishment concoction of their ideas that glossed over the most problematic aspects of their views and the result was a confusion that is still not cleared. Most writers on these two great philosophers still offer their ideas in fragmentary manners, preferring to assimilate them into the traditional background dysinformation accepted in the university. The publication in the leading philosophical press of contrasts between their ideas and received ideas will have to wait till the established leaders who had blocked their way will be no longer at the helm. In the mean time, writers have a better chance of getting published if they assimilate the ideas of Polanyi to those of kuhn and others and the ideas of Popper to those of his detractors from the "Vienna Circle".[12]

One simple example of the oversight of significant criticism should do. The idea that Newton was in error when he advocated his inverse square law is very problematic, since, clearly, this is not just any old error. On the whole, it is clear that there is a difference between errors. Some errors are due to gross negligence and are even culpable, some errors are most interesting – like Newton's. This may lead to the study of the significance of significant errors: what makes this or that error more significant than other errors? It did not. Rather, as long as this was not done, the very impression that all errors are on a par, erroneous though it is, makes it understandable that people were most reluctant to admit that Newton was in error. They did not necessarily go to study Newton's own writings, as they were more concerned with the status of his inverse square law than in his integrity. So they said, either his inverse square law is proven true, or it is true in a *façon de parler*: it is *as if* true. Consequently, some simply insist that it is proven and so true, and others explain that it is merely a *façon de parler*. Now clearly, at times science does engage in *façon de parler*: in technology simplifying

---

[11]For a discussion of the time lag in science see Agassi, J. (1975). See also Agassi, J. (1994). For a conspicuous example of the time lag see Agassi, J. (1971).

[12]For the place of Polanyi and Popper in contemporary philosophy see Agassi, J. (1981) and Agassi, J. (1988b). For attempts to assimilate Popper into the Vienna Circle, see my review of Danilo Zolo, *Reflexive Epistemology: The Philosophical Legacy of Otto Neurath*, (Agassi, 1993b).

assumptions are at time unavoidable, including such assumptions as that air or water or a piece of rubber is continuous rather than atomic. The assumption of continuity there is sheer *façon de parler*. Is this peculiar to technology? No. The counter example to this is Schrodinger's equation: his first effort was relativistic, and when he failed he moved to the Newtonian framework, *as if* Newton's theory of space and time were true. Scientists would try anything, said Einstein: they are opportunists.[13]

The very fact that scientists are opportunists, that hence some but not all research involves *façon de parler*, suffices to prove both traditional theories false. In particular, simple comparison will show that the inverse square law is taken as a *façon de parler* by technologists and literally by writers on gravity – historically and even today. How long will it take until the establishment will notice this criticism? Well, the relinquishing of both inductivism and instrumentalism is no small matter; the establishment has to be convinced that this criticism is unanswerable and to examine the major consequences from admitting it; and this takes time.

This answer seems most reasonable, yet it is a mode of deception: the establishment should admit that this is a serious criticism and advocate its examination. Instead it waits for some of its members to examine it in the hope that it be answered. This is a clear case of a bias, and one that is clearly disingenuous. When will the establishment answer this criticism of its standard procedures? When will it admit that perhaps the standards as practiced now in the commonwealth of learning invite re-examination? Doing so will be very good; it will initiate the establishment very much into a position of leadership.[14]

As long as the commonwealth of learning was marginal in western society it could be trusted to be fairly idealistic, and then it could voluntarily comply with its own more-or-less adequate standards, especially those engaged in empirical research would voluntarily comply with the standards generally observed in scientific circles. This is no longer true. Learning, especially science, is these days a matter of great prestige, integrated into society at large, and with no means to exhibit more integrity than society at large, and its leadership cannot be expected to exhibit more integrity than the leadership of other sectors of society. If this is so, then perhaps new avenues should be pursued by those not satisfied with the deterioration of scientific

---

[13] See Agassi, J. (1990b). (This appeared also as a special issue of *Poznan Studies in the Philosophy of Science and the Humanities*, **18**.)

[14] See Agassi, J. (1986b).

standards, and those who see that the traditional standards are defective anyway. Perhaps the people who are cognizant of this matter and who have the ability to develop a separate sub-society would wish to do so. For this the first task is to develop a method for sifting the grain from the chaff. That is to say, true science is to be sifted from accretions to it. This is a new wording of the problem of the demarcation of science.

3. The Demarcation of Science: Historical

The problem of the demarcation of science is a tricky business. What is required of an answer to it to be adequate? What are the adequacy criteria for it, the *desiderata* from it? One may very easily fall into a trap here. If one sticks to the traditional demand that all and only perfect theories count as scientific, then perhaps there are none, and perhaps at most only logic and mathematics will qualify. C.G. Hempel has suggested that truth is one such *desideratum*: we want of a scientific theory that it be true. This is legitimate, of course, and it reflects a worthy sentiment, but it is all the same quite unwise. Of course we want the truth. But we do not know which theory is true, yet we do know that some theories are scientific, even if their truth value is in question, and even if they are known to be false. We want science to be characterized so that some paradigm cases of science will fit the characterization and some paradigm cases of non-science will not, and, in particular, we do want Newtonian gravity as well as Einsteinian gravity to count as scientific by this characterization but not magic and astrology and parapsychology. Hence, Hempel's demand is excessive. We should not eschew it altogether though: we will stick to the *desideratum*, then, and declare the theory that answers it not scientific but a law of nature. We do not always know which theory is a law of nature and which not; but we do know that if we have assumed a theory to be one, then if it is refuted we will have to alter our view on the matter.

How then do we decide which *desiderata* to adopt for science? Worse than that. Suppose we have an answer to this question, suppose we know why we want this paradigm case and not the other. Then we also know which *desiderata* to adopt. If we know why we want to have the one paradigm case as science but not the other, then we have not only met the *desiderata* but also found the demarcation that should meet them!

Nevertheless, clearly, as we deem scientific both Newton's theory of gravity and Einstein's, we do allow false theories into the realm of science. Why? The question may be answered in different ways. We may like them both as both are useful instruments for prediction (without declaring them mere *façon de parler*). We may like them both because both are very good explanations even though not the best (= not the true ones). We may like them both as they are good exercises for students of physics. And we may find other reasons for doing so. Thus, perhaps there are different reasons for counting a theory scientific, and any theory that meets any one of them should count. This kind of policy is traditionally found highly objectionable – on the ground that each such a criterion designates an aim, and in some possible circumstances different aims may conflict. This is a general idea, integrated into decision theory, where the requirement to have one and only one aim is usually taken as self- understood. Yet this is an error. In real life we do have multitudes of aims and they often do lead into conflicts. Only then, in the light of the conflicts which we encounter, do we rethink our aims and our priorities. To demand that this should be done *a priori*, before any conflict is encountered, is to demand the perfection that is unattainable. And so confining our discourse to single aims is not perfection but excessively constraining limitation. It may be endorsed for a while, of course, but it should not mislead us to think that the excessively constraining limitation is thereby removed.

What, then, is known about the demarcation of science thus far? The matter was contested throughout the history of science. So there should be a disagreement about the very question of the scientific or non-scientific character of any theory, research activity, teaching, university department, university or any other institution, etc. In a sense this is true. Yet there is (regrettably or happily) a remarkable consensus about the *bona fide* character of a large body of theories, research activities, teachings, departments institutions, etc. This is so, I suggest, not because of any reasonable discussion of the problem and its background, but simply because there are two and only two rules that are generally admitted as inherent to science, and they prevent the possibility of the drowning of the body of science in too many lies. These are Boyle's rules and Newton's rule, and under normal conditions they function remarkably well, except that the normal conditions are not as common as they used to be.[15] It is therefore worth while to examine a few questions about them. Are they followed? How come they

---

[15] Robert Boyle, "On the Unsuccessful Experiment", in his *Certain Physiological Essays*, 1661. See also Preface to his *The Skeptical Chymist*. See also the initial Rules and Regulations of the Royal Society. These rules were proposed by the president of the Society and seconded by Boyle.

Newton's rule is presented in the "Queries" of his *Opticks*. It is well-known that this book was very popular and that the "Queries" were very influential.

work? When are they followed and when do they work? Especially if they are not sufficiently often followed, how come they still work? Do they work better when the scientific establishment observes them? What happens in areas where they ɗo not work, such as in the philosophy of science? Do philosophers of science lie more often and more significantly than scientists proper? And, finally, can these rules be improved upon?

Boyle's rules are these. First, suspend judgment about any item of non-scientific empirical evidence, and second, declare scientific only that empirical evidence that can be stated by an eye witness in court, that was repeated, and that is declared repeatable: admit as scientific the spatio-temporal generalization of a repeated item of *bona fide* observation. Boyle offered another rule, but it is much more problematic than the first two, and it is not universally observed, even though the public-relations spokespeople of science at times insist that it is. This third and problematic rule says, whenever an observation and a theory are in contradiction, prefer the observation. This second rule is not in the least new with Boyle, however. It was indeed a scholastic dictum: "*contra factum non valet argumentum*". It received a new meaning, however, by Boyle's other rules, by the institution of his idea of what counts as a fact. For this there is only a precedence – in Gilbert's report of his having repeated his experiments in front of reliable witnesses – an effort to sift the reliable from the unreliable in the large stock of available information (which he showed to be largely unreliable).[16]

Newton's rule is unprecedented, as it is a mere rider on Boyle's: when a generalization of an admitted *bona fide* observation was refuted, he ruled, it should not be merely discarded; some suitable *ad hoc* modification of it should replace it.

The philosophy of science has traditionally paid little or no attention to these two rules, even though they are central in the tradition of science. This explains the gulf between the two traditions. Anyone wishing to develop a scientifically-minded philosophy of science obviously has to take care to consider these rules, or at least not to contradict them. The first to contradict them was John Locke. Since he is the father of the new empiricist tradition and of modern conceptual analysis, and thus the father of the tradition of the philosophy of science, the breach between the two traditions was secured from the start. Since he was an admirer of science and also of impeccable credentials (being both Boyle's disciple and Newton's personal friend), the breach went unnoticed. Yet the fact of the breach is all too obvious for

---

[16]See also Preface to William Gilbert's *De Magnete* of 1600.

all to see: his empiricism takes as the starting point particular sensations, not scientific observations. The conceptual analysis that he had inaugurated was replaced by language analysis due to the works of Gottlob Frege and Bertrand Russell, as popularized by Ludwig Wittgenstein and the "Vienna Circle". His empiricism was still advocated by Russell and all of his followers except for Karl Popper, Willard van Quine and Mario Bunge. It is time to admit that Locke's empiricism and Boyle's rules clash and draw the consequences. Especially since Quine's attack on the kind of empiricism that Locke has inaugurated is considered for decades a classic ("Two Dogmas of Empiricism"), it is time to do that.

The difference between Boyle's rules and Locke's analysis was noticed by Immanuel Kant, though he did not state it explicitly. Kant declared sensations private, and so unscientific. He declared the wording of these sensations as universal statements within the proper theoretical system proven and so scientific. This is a sufficiently close wording to Boyle's wording of his rules, perhaps, but certainly it makes Newton's rule redundant.

The reasons Locke began the tradition of the analysis of sensation is that the senses are known to mislead. Francis Bacon, the great trailblazer of modern philosophy of science, discussed repeatedly the fact that the senses mislead, but he said no more about this matter than that the unprejudiced do not see facts wrongly. The errors which are due to mis-perceptions, are always due to misjudgments, to false theories, he said. He called them "prejudices of the senses" and "prejudices of the intellect" respectively, and he declared the one sort to be a version of the other. He also promised to develop safeguards against mis-perceptions but did not. Evidently he was himself in error, as he reported having observed magical cures and as he declared Copernicus a charlatan.[17] Locke deemed it essential to go back to the details of perception in order to show that we do not see the motion of the sun: we do not see motion at all, he declared; we always infer it. Berkeley continued this line of thought, but destructively: we do not see substance, and so our theories can never logically impose on those who endorse all observations of fact the theory that substance exists. Hume did the same for causes, except that he was doing this in defence of science, relying, as he did, on Newton's famous declaration that he did not know the cause of gravity and was not engaged in feigning hypotheses.[18] It is clear today that to

---

[17]For Bacon's ideas and their significance see Agassi, J. (1988c). See also Agassi, J. (1989), which is a review of the book by Peter Urbach on Bacon, and my review of Charles Whitney, *Francis Bacon and Modernity, ibidem*, 223-5.

[18]See Agassi, J. (1985).

reconstruct observations from sensations is a complex matter, depending on highly sophisticated theories. This raises a serious difficulty: if observations depend on theories, but their authentication itself rests on theories, then this authentication is thereby rendered questionable.

This is definitely the case. This leads to the question, can science be authenticated in any other way? If not, can it be carried out with no authentication? The affirmative answer to the first question is still contested. The affirmative answer to the second question must be admitted as an observation: though no authentication of science was ever generally endorsed, the venture still goes on; nor do individuals who deny that authentication can ever be found barred from contribution to scientific research. This being so, the rationale for Locke's analysis of sensations is lost, and the failure of that analysis is thus inconsequential for science, as is the failure of the same analysis when performed by Ernst Mach, Bertrand Russell or Rudolf Carnap. Quine's objection to it is not hostile to science.

Boyle's rules are violated not only by all reconstructions of observations from sensations by the analysis of sensations; they are also violated by all theories of confirmation of generalizations by particular observations (such as those of Carnap and of Hempel). By Boyle's rules particular observations are to be left alone, neither endorsed nor rejected, unless and until their generalizations are endorsed in accord with standard procedures.

Neither Boyle's rules Nor Newton's speak about authentication or validation of theories. It may be remembered that Boyle also demanded that when a theory and a generalized observation clash, priority should be given to the observation: it is the theory that has to be rejected. One should not rush to the conclusion that Boyle admitted that scientific hypotheses are refutable; he explicitly declared the excellence of the mechanical hypothesis to lie in its irrefutability. He only demanded that observations should never be rejected on the basis of a theory. This is consistent with the famous Duhem-Quine thesis that there is no reason to prefer the rejection of one (general) theory over another when some rejection has to be effected: both Boyle's rules and the Duhem-Quine thesis demand to hold fast to the (generalized) observation. The demand of logic, when the observation contradicts a theory, is merely to remove the contradiction. Pierre Duhem, truth to tell, was ready to temper with the (generalized) observation itself somewhat, but this, at most, is contrary to Newton's rule that demands to temper with it only when it is refuted.

Both Boyle's and Newton's rules are problematic. Yet there is almost no discussion of them in the literature, scientific, philosophical or sociological.

This is disquieting. That we do not quite know what is a *bona fide* eyewitness testimony is a fact, though we do know that expert witness testimony is not it. Thus, that witch hunting was conducted in courts, and was in full swing while Boyle's rule was instituted, is an unquestioned historical fact. Boyle himself was very concerned about the status of his own rules, and there is historical evidence that the matter did disturb a number of spokespeople for the new scientific society, but the doubts were quelled and the rules were deemed *sine qua non* for science proper.

The question of the demarcation of science may be pursued either while accepting Boyle's and Newton's rules or while modifying them explicitly; ignoring them will perpetuate the cleavage between the scientific tradition and that of the philosophy of science.

4. The Demarcation of Science: Recent

The idea that science is certitude remained the official view despite all tremors that science regularly underwent, but the overthrow of Newtonian mechanics was the last blow. Something had to be acknowledged. As with all cases of failure, the response can be either that efforts in the same direction must be redoubled or that the target is unattainable and aught to be downgraded or even replaced. The idea of redoubling the effort was made by both Ludwig Wittgenstein and by P.W. Bridgman. Both offered suggestions as to the securing of the attainment of the goal of empirical proof in science. Both failed, and even for reasons that were easily predictable: there was no ground for their supposition that the increases in precision of techniques which either of them had recommended was the right one. The alternatives were surrogate certitude. Each of the surrogates was at least as problematic as the original, as there was no more guarantee that the surrogate would work, nor that it was the right surrogate. The most common surrogates were probability and relative truth. Now probability assessments were problematic: how are we to know that they are true or at least probable? Relative truth is relative to some initial suppositions that may be remote from the truth. To this relativists respond saying, there is no absolute truth. That response renders utterly separate each system relative to which certain ideas are supposedly true or false, and if one statement happens to be true in one system and false in another, there is nothing that can be said or done about it. So if in one system an empirical statement is true and in another it is false, the choice of the system rather than the observed facts determine matters. This sounds strange, but the Duhem- Quine thesis insures the possibility of such a case, and the conjugate Duhem-Quine radical untranslatability thesis renders doubtful even the claim that the same statement refers to the same factual situation.[19]

As to the problem of demarcation, the surrogate certitude thesis replaces the view that science is certitude with the view that science is surrogate certitude, whatever the surrogate is. The idea of probability as a surrogate certitude leads to the demarcation of science as probability, meaning high probability, with a proviso as to what degree is sufficiently high. And the theory of relative truth declares scientific all ideas certain given the framework, whatever the framework is, which makes the framework itself science at once, even if it is theology or a superstition of the weirdest kind. To insure that this is not so one may take as the framework Newtonian mechanics or Einsteinian one, as Kuhn has recommended. But why?

If we take Boyle's rules as the determining of what is empirical data, and offer as a rationale for them the claim that they are required to insure critical examination of the data, then we may suggest a view as to what is the motive and the task of the scientific endeavor. Suppose the motive is curiosity and the task is explanation (as Boyle suggested), and the question then is what is an explanation and what is an adequate explanation? A partial answer to this question is Popper's: an adequate explanation must be testable, i.e., refutable-in-principle, i.e., open to refutation if it is false, i.e., if it is false there is a chance that those who wish to refute it will succeed.[20]

There is a body of literature critical of Popper's proposal. Most of it is criticism aimed at showing that according to Popper even refutation is not certain. This is true, but why it is taken as criticism of Popper the critics do not say, unless they mean, if there is no full empirical verification, at least they demand to secure empirical refutation. But Popper takes Boyle's rules as correct (and as the rationale behind them he takes the ideas described in the previous paragraph), and so he has neither the wish nor the need to answer that demand. Other comments on Popper's proposal, mine included, are matters of fine tuning.

My critique of Popper is that the characterization or criterion he has offered is partial: refutability is empirical character, and whereas all science is empirical (ever since mathematics was excluded from it, somehow arbitrarily

---

[19]See Agassi, J. (1992).

[20]The *locus classicus* of Popper's demarcation of science still is his *Logik der Forschung* of 1935, English version, *Logic of Scientific Discovery*, 1959 – both available in many later editions – though he has published additional interesting material on the matter in his *Conjectures and Refutations* of 1961, *Objective Knowledge* of 1972, Chapter 1 and elsewhere.

perhaps), obviously not every empirical research is scientific. The same goes
for backing or positive result or success, which Popper takes to be always
the same as empirical corroboration, namely the passing of a (severe) test:
whereas, clearly, every empirical corroboration is success, not all success is
corroboration. It is one thing to say what makes the empirical success empirical and another to say what makes the empirical success success. Any
claim for empirical success is a claim for refutability of the success, but the
kind of the success under discussion depends on its context, on the task at
hand: what makes anything success is its claim to be a fulfillment of that
task. In theoretical science it is explanation, in experimental science it is
the design of a test, in technology it is the compliance with the regulations
of the law of the land.[21]

To see how much better it is to leave open the question, what is the
task at hand, take a partial task. If the task is testable explanation and
this is very hard to come by, then success is achieved even by finding an
explanation that is hopefully but not obviously testable, so that the success
of a hypothesis, the claim that it is backed, is displayed by the very exercise
that exhibits its explanatory power. This can be sharpened: at times the
new explanation is *a priori* known to be false, yet it is greeted as a great
success. (Here Popper's theory of all success as corroboration is refuted since
by his rule the degree of corroboration of an already refuted hypothesis is
the lowest possible and cannot be raised.) This is the case, for example, with
Niels Bohr's model of the atom, that fits at best only the first column of the
table of elements. Its success was judged differently, and it did not wait for
any empirical test. Alternatively, the offer of a testable theory, even if it is
not yet explanatory, may count as a success. This is the case, for example,
of the Bohr-Kramers-Slater theory of 1924, which was a success when it
was proposed, just because it awaited empirical test. Of course, a testable
theory should hopefully be made explanatory if it is empirically tested with
positive results or even merely encouraging ones; not all presentations of
testable non-explanatory theories are successful, however. The case of the
Bohr-Kramers-Slater theory was a failure, and a most significant one, since
its originators ware embarking on a project of developing an explanatory
theory which takes the conservation of energy to be statistical, and a whole
bunch of such theories was refuted *en bloc*. This illustrates yet again the

---

[21] For my critique of Popper's demarcation of science see Agassi, J. (1991a), Popper's
Demarcation of Science Refuted, 1-7. For the diversity of aims in science see my "Pluralism
and Science", *ibid.*, 99-119.

ancient (Socratic) maxim, usually neglected, that refutation is enlightening. William Whewell and Karl Popper are the modern thinkers who built their theories of success on that maxim; yet they took success to be of one kind. Even the success of the one kind, corroboration, can be partial, as Nancy Cartwright observes: *A theory is lucky if it gets some of the results right some of the time.* (Of course, literally speaking, a theory cannot be lucky any more than it can lie.)[22]

The thrust of the exercise in the previous paragraph is to show that the *desiderata* for a criterion of demarcation are not a mere instrument to tackle the problem of specifying the criterion but are even the very criterion itself. The oversight of this is the confusion between a *desideratum* and a touchstone. The touchstone is a useful criterion whose connection with the *desiderata* is tenuous. What makes gold desirable may be its resistance to corrosion and its malleability, and these can be used as criteria for testing the claim that a given object is made of gold, such as dropping a drop of acid on it or biting it. The touchstone, the piece of stone that scratches gold, does so for reasons that may be unknown, and it has nothing to do with the desirability of gold. Touchstones may be handy, but they carry no intellectual weight. A touchstone for scientific character is often such a thing as the presence of mathematical formulae. This invite forgeries, of course. Applicability is a very good touchstone, but one has to observe that many *bona fide* scientific theories have no known applications and many highly applicable theories are mathematical and/or technological, with no scientific interest (as yet). Also, one has to observe that not all claims for applicability are scientific: those are that the law specifies the severity of the test involved in the validation of their claims for applicability.

5. Observed Regularities and Laws of Nature

There is something otiose these days about the very preoccupation with the problem of demarcation of science, akin to the problem of identifying gold: it is not sufficiently disinterested. Of course, while engaged in research on gold chemists may be as oblivious to its market value as is reasonable to expect of them, yet most people are more concerned with wealth than with the properties of gold. The same holds for science, and more abjectly so. And so the demarcation is in order of the inquisitive, disinterested concern with the problem of demarcation of science from the superficial wish to serve some self- interest by claim to be scientific. This, however, is very repugnant,

---

[22]See Wettersten, J. and Agassi, J. (1991).

as it seems to be an invitation to motivation and to sou˚ searching. Of course, the point is that we should distrust easy touchstones and be a bit critical about matters, and, most specifically, not take science to be identical with its material success, at lest not as a matter of course. This is the point of failure of most studies of the problem of the demarcation of science and of the problem of induction which is derivative of it.[23]

There are diverse ways to insure the avoidance of the crude identification of science with scientific success. One is to speak of noble scientific failures, such as those Planck and Einstein met for decades, and of great scientific difficulties, past and present. One may also discuss the material value of science as such: does it always improve the quality of life? This is an open question ever since it was empirically shown that it can very easily do the opposite by being put to military uses, by causing pollution, nuclear and other, and by raising the level of poverty and of population explosion in parts of the world where its benefits are less vigorous than in the West.

But one can likewise sing the praise of obviously not scientific studies, and by critically examining the intellectual significance of science. For example, one may examine the rules of science, Boyle's, Newton's, or any other. Why should we concentrate on repeatable events alone? Are astronomical events repeatedly observable? What exactly is the claim for their repeatability? I will not examine these questions here. Rather, let me return to the rationale of repeatability. The official one offered by Boyle, Popper, and Mario Bunge, is that we want our empirical record testable. This is debatable, but I will not debate it. Rather I will say, it is not meant to exclude historical studies proper, and historical records are the paradigms of records of non-repeatable events. Yet there is a reason to prefer repeatable events just because they are repeatable. It is this.

Life depends on regularities: it is an empirically repeatedly observed fact that regularities are essential to life. This is expressed in the very repeatability of certain events, of course. And so we wish to preserve the regularities that are life-sustaining. Is this necessary? Yes: some regularities fail and this creates disasters. We wish to avoid those. How?

The juvenile literature teaches us of the benevolence of the laws of nature. We do not like friction; it robs us of the scarce resources of energy that are at our service. Yet imagine a world without friction: it is a veritable nightmare. Now this exercise is thought provoking, and it is preformed regularly by all

---

[23] For a discussion of criteria and touchstones see my "Questions of Science and Metaphysics", reprinted in: Agassi, J. (1975).

sorts of ingenious science-fiction writers. But taking it literally is foolish and even dangerous, and its popular dangerous expression these days is the gaya hypothesis which says that Mother Earth takes good care of Herself and so we need no' worry overmuch about total destruction.

Though worry is neither here nor there, what is in order here is, if not responsibility, then at least curiosity. We do observe regularities. It is a matter of regularly observed fact that we do. The observed regularities are *a priori* either stamps of the laws of nature or mere local characteristics. Which is which? We do not know, but we do possess a few simple and fairly commonsense examples, beginning with the one known to all: the seasons and the seasonal precipitation are regularities, the one reliable, the other not. Comes science and tells us that the seasons, too, are unreliable, ephemeral, dependent on local configurations. Of course, this change is achieved by the rupture of the sense of the local. It is this rupture that made modern science a shock to our system, one that we have not begun to overcome. The laws of nature are truly universal, and the idea of true universality is awesome. To take one of the latest examples, we are convinced today with no hesitation that oxygen and water are local; almost everywhere in the universe they are sufficiently absent to make life there impossible unless managed with adequate life- support systems.[24]

The question, thus, is fairly tricky: which regularities are local and which are thanks to the laws of nature? We do not know as we do not know the laws of nature. What we do know is that we get a shock every time we manage to refute what we had taken for granted as the true, truly universal laws of nature. The shock at times leads to a refusal to recognize the facts. But this is ostrich policy, of course. So we have to come to terms with the fact that the universe is a hostile place, and that we do not know if its hostility is due to the brutality of the laws of nature or a mere local characteristic, however widely spread. After all, we are here, and by the grace of the laws of nature for sure. So do not blame the laws of nature. They do not work for us exactly, but not against us either – not necessarily so. Things may still depend on us.

References

Agassi, J. (1963), *Towards a Historiography of Science, History and Theory, Beiheft*, **2**; facsimile reprint, Wesleyan University Press, 1967.

---

[24]See my review of "The Brundtland Report", Agassi, J. (1991b). See also Agassi, J. (1990c).

Agassi, J. (1971), *Faraday as a Natural Philosopher*, Chicago University Press.

Agassi, J. (1975), *Science in Flux, Boston Studies in the Philosophy of Science,* **28**.

Agassi, J. (1977), *Towards a Rational Philosophical Anthropology*, Kluwer.

Agassi, J. (1981), *Science and Society: Studies in the Sociology of Science, Boston Studies,* **65**.

Agassi, J. (1985a), *Technology: Philosophical and Social Aspects*, Kluwer, Dordrecht and Boston.

Agassi, J. (1985b), The Unity of Hume's Thought, in: *Hume Studies,* **10**, Supplement, 87-109.

Agassi, J. (1986a), The Politics of Science, *J. Applied Philosophy,***3**, 35-48.

Agassi, J. (1986b), Scientific Leadership, in: Carl F. Graumann and Serge Moscovici, (ed), *Changing Conceptions of Leadership*, Springer, NY, 223-39.

Agassi, J. (1987), The Uniqueness of Scientific Technology, *Methodology and Science* , **20**, 8-24.

Agassi, J. (1988a), The Future of Big Science, *J. Applied Philos.*, **5**, 17-26.

Agassi, J. (1988b), *The Gentle Art of Philosophical Polemics*, Open Court, LaSalle IL.

Agassi, J. (1988c), The Riddle of Bacon, *Studies in Early Modern Philosophy,* **2**, 103-136.

Agassi, J. (1989), The Lark and the Tortoise, *Philosophy of the Social Sciences,* **19**, 89-94.

Agassi, J. (1990/91), *The Siblinghood of Humanity: Introduction to Philosophy*, Second edition, Caravan Press, Delmar NY.

Agassi, J. (1990a), Newtonianism Before and After the Einsteinian Revolution, in: Frank Durham and Robert D. Purrington, (eds), *Some Truer Method: Reflections on the Heritage of Newton*, Columbia UP, NY, 145-176.

Agassi, J. (1990b), Ontology and Its Discontents, in: Paul Weingartner and Georg Dorn, *Studies in Bunge's Treatize*, Rodopi, Amsterdam, 105-122.

Agassi, J. (1990c), Global Responsibility, *J. Applied Phil.*, **7**, 217-221.

Agassi, J. (1991a), The Ivory Tower and the Seats of Power, *Methodology and Science,* **24**, p. 64-78.

Agassi, J. (1991b), *World Commission on Environment and Development* in: *International Review of Sociology, Monographic Series,* **3**, Borla, Roma, 213-226.

Agassi, J. (1992), False Prophecy versus True Quest: A Modest Challenge to Contemporary Relativists, *Philosophy of the Social Sciences*, **22**, 285-312.

Agassi, J. (1993a), The Philosophy of (ptimism and Pessimism", in C.C. Gould and R.S. Cohen, (eds), *Artefacts, Representations and Social Practices. Essays for Marx Wartofsky, Boston Studies in the Philosophy of Science*, 349-59.

Agassi, J. (1993b), Neurath in Retrospect, Iyyun:eCIT, **42**, 1993, 443-453.

Agassi, J. (1994), Minimal Criteria for Intellectual Progress, *Iyyun*, **43**, 61-83.

Agassi, J. (1995a), Contemporary Philosophy of Science as a Thinly Masked Antidemocratic Apologetics, in: the *Robert Cohen Festschrift, Boston Studies in the Philosophy of Science*, forthcoming.

Agassi, J. (1995b), The Philosophy of Science Today, in: S. Shanker and G.H.R. Parkinson, (eds), *Routledge History of Philosophy*, Vol. 9, forthcoming.

Lakatos, I. (1963), Proofs and Refutations, in: *The British Journal for the Philosophy of Science* 14, pp. 1-25, 120-139, 221-243, 269-342.

Laor, N. (1985), Prometheus the Impostor, *Brit. Med. J*, **290**, 681-4.

Popper, K. (1961), Conjectures and Refutations.

Wettersten, J. and Agassi, J. (1991), Whewell's Problematical Heritage, in: M. Fisch and S. Schaffer, *William Whewell: A Composite Portrait*, Oxford UP, 345-69.

BIOGRAPHICAL NOTES

Diederik Aerts is Senior Research Assistant of the Belgian National Fund for Scientific Research at the Theoretical Physics Department of the Free University of Brussels. He is a member of the Council of the International Quantum Structures Association, and of the Council of 'Worldviews'. he is also coordinator of CLEA, an interdisciplinary research center at the University of Brussels.

Sven Aerts is Ph.D. student in CLEA at the Free University of Brussels.

Joseph Agassi, Professor of Philosophy, Tel-Aviv University and York University, Toronto. M.Sc from Jerusalem; Ph.D. from London School of Economics. Fellow, American Association for the Advancement of Science; Fellow, Royal Society of Canada; Fellow, World Academy of Art and Science.

Major books: *Towards an Historiography of Science, History and Theory*, Beiheft 2, 1963; 1967; *The Continuing Revolution, A History of Physics From the Greeks to Einstein*, 1968; *Faraday as a Natural Philosopher*, Chicago UP, 1971; *Science in Flux*, Boston Studies, 28, 1975; *Paranoia: A Study in Diagnosis* (with Yehuda Fried), Boston Studies, 50, 1976; *Towards a Rational Philosophical Anthropology*, Kluwer, 1977; *Science and Society: Studies in the Sociology of Science*, Boston Studies, 65, 1981; *Psychiatry as Medicine* (with Yehuda Fried), Kluwer, 1983; *Technology: Philosophical and Social Aspects*, Kluwer, 1985; *The Gentle Art of Philosophical Polemics: Selected Reviews*, Open Court, 1988; *Diagnosis: Philosophical and Medical Perspectives* (with Nathaniel Laor), Kluwer, 1990; *The Siblinghood of Humanity: Introduction to Philosophy*, Caravan Press, Delmar NY, 1990, 1991; *Radiation Theory and the Quantum Revolution*, Birkhauser, Basel, 1993; *A Philosopher's Apprentice: In Karl Popper's Workshop*, Editions Rodopi, Amsterdam, 1993.

Arne Collen has provided international research consulting services to students, colleagues, and organizations since 1971. His consulting work, invited lectures, and short courses have taken him to Austria, Canada, Denmark, Fiji Islands, Finland, France, Germany, Italy, Mexico, Poland, Russia, Sweden, Ukraine, and the United States.

Through contracting for individual consultation and team projects, he has assisted others to: 1) carry out all aspects of the inquiry cycle, 2) select and use research methods relevant to studying and working with human beings, and 3) construct research methodologies for human science inquiry.

He also serves in such a capacity as Full Professor of Human Science, Psychology, and Systems Inquiry at Saybrook Institute, located in San Francisco, California, where he works in a supervisory role to mid–career professionals during their doctoral graduate studies.

Biography: B.A. in Psychology 1965 from San Francisco State University; M.A. 1968 and Ph. D. 1971 in Psychology from Ohio State University. Assistant Professor 1971–1975 Marshall University. Academic Dean 1976–1978 California School of Professional Psychology. Associate Professor 1978–1983, Professor 1983–present Saybrook Institute. Research/organizational consultant, methodologist.

David Freedman received his B.Sc. from McGill and his Ph.D. from Princeton. He is Professor of Statistics at the University of California, Berkeley, and has been Chairman of the Department. He has published several books and many papers in probability theory and statistics. His current research interests are in the foundations of statistics and policy analysis. He has worked as a consultant to the World Health Organization, the Bank of Canada, and the U.S. Department of Justice.

Maria Carla Galavotti is Professor of Philosophy of Science at the University of Trieste. She has worked on the foundations of probability, the nature of statistical methods and their application to empirical sciences, and problems like explanation, prediction, causality. Among her publications is a book on probabilistic explanation: *Spiegazioni probabilistiche: un dibattito aperto* (Bologna: CLUEB, 1984); the issue of Topoi (IX (1990), n. 2) on "Recent Developments on Explanation and Causality" and the volume *Probabilità, induzione, metodo statistico* (Bologna: CLUEB, 1992). She has also edited a volume of *Erkenntnis* (XXXI (1989), n. 2-3) on "Bruno de Finetti's Philosophy of Probability" together with R.C. Jeffrey, a volume of *Theoria* (LVII (1991), n. 3) on "The Philosophy of F.P. Ramsey" and a collection of F.P. Ramsey's unpublished papers under the title *Notes on Philosophy, Probability and Mathematics* (Naples: Bibliopolis, 1992).

Paul Humphreys is Professor of Philosophy at the University of Virginia. He was educated at the University of Sussex and Stanford University, studying logic, physics, statistics, and philosophy. His principal interests lie in causation and explanation, probability and statistics, and more recently, computational science. He is an editor of Synthese and of Foundations of Science, the author of *The Chances of Explanation* (Princeton, 1989) and edited the recent three volume collection Patrick Suppes: Scientific Philosopher.

Patrick Suppes is Lucie Stern Professor of Philosophy (Emeritus) at Stanford University. He earned his Ph.D. at Columbia University, studying with Ernest Nagel, and has since held concurrent appointments at Stanford in statistics, psychology, and education, as well as in philosophy. He has worked in a wide variety of areas, including theory structure, measurement theory, foundations of probability, causality, experimental and theoretical psychology, and robotics. He is a member of the U.S. National Academy of Sciences and was awarded the National Medal of Science in 1990.

Bas van Fraassen is Professor of Philosophy at Princeton University. He earned his Ph.D. at the University of Pittsburgh, and has taught at the University of Toronto and the University of Southern California before joining the Princeton faculty. His principal publications are *The Scientific Image, Laws and Symmetry, Quantum Mechanics*, and *An Introduction to the Philosophy of Space and Time.* He has also published in the areas of philosophical logic, probability theory, and theory structure. He is a past President of the Philosophy of Science Association and a founding editor of The Journal of Philosophical Logic.

On Association for the Foundation of Science, Language and Cognition, AFOS;
some history and some recent developments

The initiative to form AFOS emerged in March 1993 and since then it is rapidly gaining momentum. At present it is supported by almost three hundred scholars from all over the world.

The objectives of AFOS, as described in the Constitution, is to promote investigation of foundational and methodological issues in the sciences, language, and cognitive studies. 'Sciences' should be construed here to include both the natural and the human sciences. AFOS subscribes to the view that this is best done by adhering to the highest standards of logical, scientific and philosophical rigour. In order to promote such investigation, AFOS seeks to support activities such as, but not limited to: publication of scholarly journal(s) and books in the field; sponsorship of scholarly meetings, and fostering communication between its members.

The AFOS Coordinating Committee and regional representatives are: Joseph Agassi (Israel–Canada), Newton C. A. da Costa (Brazil), Dennis Dieks (Netherlands), Irina S. Dobronravova (Ukraine), Leo Esakia (Georgia), Paul Humphreys (USA), Javier Echeverria (Spain), Maria Carla Galavotti (Italy), M. Kaiser (Norway), S. P. Kurdyumov (Russia), B. C. van Fraassen (USA), Marco Panza (France), Patrick Suppes (USA), Ladislav Tondl (Czech Republic), Ryszard Wójcicki (Poland), Jan Zytkow (USA). The Coordinating Committee will terminate its activity as soon as the Executive Committee is elected.

Further information and access to the initiative. Full membership shall be open to qualified scientists, theorists of science, and philosophers who are in sympathy with the objectives of the Association. For further information contact **Ryszard Wójcicki**, Institute of Philosophy and Sociology PAN, Nowy Świat 72, 00–330 Warszawa, Poland, Fax:(48)-22-267823), E-mail: iandi@plearn.edu.pl or any member of the Coordinating Committee .

'94 AFOS Workshop on Foundations of Science was held at Mądralin (the vicinity of Warsaw, Poland) on August 15 – 26 and then followed by the '94 Conference of the International Union for History and Philosophy of Sciences (Warsaw, August 26 – 29). The Proceedings of the two meetings will be published by **Rodopi** as Vol. 44 of the *Poznań Studies in the Philosophy of the Sciences and the Humanities*. For a summary of the workshop see the next piece in this volume.

An affiliated meeting of AFOS will be held at the next Congress of Logic, Methodology and Philosophy of Science, Florence (Italy), August 19 – 25, 1995.

Foundations of Science
1 (1995/96), 161-166

Joseph Agassi

SUMMARY OF AFOS WORKSHOP, 1994

It is a great honor to be invited to sum up the proceedings of this, first AFOS workshop. I do not quite know how to do it. It is difficult enough to work within a wel-established framework, but I do not think there is an established framework for this. One of the most important philosophers of science of the century, Michael Polanyi, said that science is a system of workshops, since the master scientist, like the master artist, does not know what exactly makes his workshop what it is, as his knowledge is tacit, unspoken. Certainly there is tacit knowledge in every activity, but for my part it seems to me that the workshop mentality is the opposite of what Polanyi describes, as it is fluid, and as the leadership of the proceeding keeps changing so that junior members of a workshop can have their day and make a contribution to the goings on, as the agenda can alter as a result of the interaction of memebers, of the group dynamics. This, I think is the spirit of science, and it is described in a few works where some historical workshops are described. Let me mention two studies that reflect it, Robert Jungk, Brighter than a Thousand Suns, and Einstein and the Generation of Science by Lewis Feuer. The main point to stress here is that the group dynamics is of great significance, and whatever it is it must be in friendly spirits. I think I need not say more on this as to our workshop: the friendly spirits were evident, and the procedures were developed satisfactorily as we went along. Since the discussions centered around papers, let me say that some of the papers were unpanned, on which more soon, and that some of the members here contributed significantly even though the did not read papers. I refer especially to the senior members of the workshop. I know that senior people do not particularly like to be referred to as seniors, so let me mention here Bertrand Russell's terrific "How to Grow Old", wher he says, seniors can stay young by staying involved in the problems of the young. In this sense we are all young here.

The main things to say about our workshop is, first, that it took place in terrific surroundings, in ideal conditions – of isolation plus excursions. The firs task I have, therefore, is to express the gratitude of the membership to the donors, the organizers and the staff. Second, the workshop takes place under the auspices of AFOS. We can all be proud of sharing the privilege of being members of the first AFOS workshop, thus contributing significantly to a new and significant tradition. We have devoted a significant session to the aims and character of AFOS and we agreed that it maintains the scientific attitude and strives for a scientific world-view, and that it should have the unusual task of maintaining high standards while staying as open as possible. As to openness, we agreed that there is no room in our midst to obscurantism, even of the fashionable sort, though we will not close our gates to obscurantists who are willing and able to have interesting dialogues with us. As to the maintenance of high standards, the most obvious way of trying to do so is by writing on one's gatepost what Plato wrote on the one to his Academy: no entry for those not expert in mathematics. By demanding ever more bacground knowledge we may try to raise our standards. This was tried in the scientific v rld f the Post-World-War-II period, when nuclear physics became so prestigious and lucrative and the model of the oxymoron of combined scientific character and secrecy. We want to start differently, requiring high standards but no specialized konwledge.

How exactly is not clear, and our little workshop was a minor experiment in this respect. We tried to be explicit, to procure explanation as much as we could of whatever item of background technical information we were using. I do not wish to present an idealized image of the workshop. There was generally too much air of familiarity with some rich background material. Too many times we mentioned some texts as such that we all are familiar with and approve of, especially some of Prigogine's popular works and Rene Thom's. Had we had more time and more ability, we would institute talks on these writers of the kind we had about chaos. There is room for improvement, and the next workshop will be better, we all hope.

There was, particularly, trouble with the contingency that wished to speak of synergetics. Most of us did not and still do not know what they are saying, despite the fact we were also given some published material to help us with the task. It is a cause for pride that the situation was taken as a challenge and that youn members of the workshop volunteered to present the essentials of the theory of fractals, of self-similar functions and of the main self-similar functions that are solutions to some non-linear differential equations that are central to the theory of chaos. There were a few short impromptu expositions, and these were excellent; unfortunately, some of us were too shy to participate as actively as workshops require.

Te efforts to understand the basic of the mathematics involved should have helped us understand what our synergeticists were trying to say, and, indeed, when the next talk on synergetics was given the speaker referred to the background material as helpful, but after covering some familiar background material he went over to the material specific to that group then the others lost him. Pity.

In addition to the workshop on synergetics, to which I will return last, we had a few more workshops. We had one, of Adam Grobler,'on the views on science and on rationality of Larry Laudan as an effort to improve upon other, better known ideas on the topic, and on the shortcomings of these ideas and on how these may perhaps be overcome. As this discussion was rather abstract it was wisely complemented by a broad outline of the history of physics in the last three or four centuries. I will not try to summarize it as it was already to concise for my taste. We also had a workshop on data and phenomena with a special reference to the history of the discovery of the phenomenon of continental drift. I cannot avoid mention the terrifically clear and concise and interesting presentation by Matthias Kaiser of that chapter in history.

We had a workshop on the social background of science and its interactions with the methodology of science, a topic also covered in the already mentioned discussion of phenomena. We had a lively discussion of mechanized heuristic, of the sociology of science and its strong interaction with the development of science in accord with chaos theory, of the ethics and the methods of science, and of both the Warsaw-Lvov school of logic and the Vienna Circle each in its peculiar social settings. We heard a trailblazing description by Sven Aerts of measurement interacting with the measured not only in quantum mechanics, but also in classical mechanics and in social science.

Finally we had two talks of Ryszard Wójcicki, one by Paul Humpreys and one by Mauricio Suarez. The last two, this afternoon, were on Gierie on realism that I will not summarize for want of time despite the lively debate it generated; the latter was on computational empiricism that depicted the changes that modern powerful automatic computational methods have introduced into the methods practiced in science. It certainly was new to me and I found it refreshing and exciting.

Let me discuss Ryszard's papers a bit, as he was not clear about his aims in presenting them. The first was his own version of the history of the theory of science in the twentieth century. He took off from the views of the Vienna Circle of science and he claimed that they were concerned with the problem of demarcation of science, that their researches on this were failures, though important ones nonetheless, and that they were superseded by the new theories that take into account when discussing science, its history, the psychology and the sociology of scientist, and so on. He also said that the Vienna Circle was important despite its failure, but he did not explain. He also included Popper in that circle, which is standard if to my mind unacceptable.

I find this disturbing. The Vienna Circle saw itself as a crusaders against metaphysics, and this should be mentioned in respect for people's opinions of what they do even we do not share these opinions. Also, the Circle was concerned not only with the problem of demarcation of science, which, I agree with Ryszard, is not the most important, but also with the problem of induction which still is with us: how do we gain theoretical knowledge from experience? Popper's answer, following Einstein, is, by empirical criticism. This answer, which seems to me to be central to the progress of methodology in our century, was ignored by the whole profession, as was Popper himself. He was first recognized officially by scientists, not by philosophers, and these still have the problem of what to do about past injustices to him. They try to gloss over this question, but they will fail.

Ryszard's second paper is a presentation of a new, semi-formal theory of semantics, i.e. of meaning and truth. The background for this is a dissatisfaction with the classic, standard theory of Tarski, and I cannot go into this now. The paper displays a very interesting and very novel approach, in which, contrary to Russel and the whole of the analytic tradition, but utilizing vaguely the idea of possible worlds, facts can be said to obtain or not to obtain, where these decide respectively the truth or falsity of their descrition. If this essays is not found at once grossly defective, and I for one cannot predict on this matter, then it is going to be cited a lot and for a long time, and we will say, I was there when it was first read.

I feel the need to say more about the workshop on synergetics just because we all felt so frustrated about it, the advocates of the ideas of synergetics who come from the former Soviet Union and the rest of us. Frustration should be taken as a challenge as long as there is good will, especially since, as was noticed, the frustration is in part at least the result of their isolation that is now over and this invites taking advantage of the new contacts. Let me say a few words on the matter at hand before bringing this summary to a close and opening the workshop's floor for discussion.

Synergetics is a collection of ideas and research techniques related to chaos theory, guided by the hope that the new mathematical technique will help progress in efforts to overcome the standard problems associated with self-regulating systems that were left unsolved by the older cybernetic technics. Somehow the feelin was conveyed that the spokespeople for synergetics here felt vindicated every time an assertion from cynergetics or from chaos theory was conceded. Once there was even the impression that our synergeticists even claimed credit for the fact that some physicists have invented a maser apparatus of which they received Lenin prize. This may be just an impression, and perhaps not all of us share it, but we owe it to our friends to tell them that some of us did share it and that they will be better off trying to prevent giving such impressions. But let me speak of the maser apparatus itself.

Our synergeticists stand out also in that they represent clinics that use the apparatus in question to cure a variety of illness in a method that is a variant of the traditional Chinese acupuncture method that I will not discuss. Two claims are made relative to this technique or set of techniques that they advocate. First, that it is an expression of a new theoretical synergetic breakthrough, and second that it works. It is clear to me that on the assumption that it works there is no theoretical breakthrough here, much less a physics of the alive and a quantum medicine. These are just fancy names. the basis for this is the idea of quantum or non-quantum resonance. Now a resonance is a localized phenomenon, like the sound on the radio that is heard only when its dial is at a certain position, but not all localized phenomena are resonance, for example, pain. The use of the word "resonance" here is sheer metaphor. In the first lecture here on the topic we were told that the word "non-linear" is used in a sense non identical with that in the expression "non-linear differential equations" Let me stress that I have no objection to metaphor – only to pretense that a metaphor carries more information than it does.

Let me say, then, a word on the question of the work of the clinics they representant as alternative medicine. I think alternative medicine serves us an important function, and it is regrettable that it is not better presented and developed than it is. Established medicine is too powerful and it is in need of public control and checks, an alternative medicine is the natural system to prompt this control. Though established medicine began in earnest with the development of hygiene and nutrition which are "soft", the tendency of established medicine is towards preferring "hard" techniques. No doubt, physical exercises and proper diet are better than open-heart operations if and when they achieve the same results, yet usually they come too late for patients who are carted to the operation theater. Yet there is no doubt that in matter of giving birth obstetricians have earned the sever censure that the received from the feminist movement for their hard techniques, especially for their unnecessary use of cesarean sections for patients who are properly insured. There is even the matter of surgeons preferring to date mastectomy over lumpectomy when at best the advantage of the severe method over the less severe one is still not empirically corroborated. There is also the matter of over-diagnosis that the literature provided to us by our synergeticists mentions. All in all, the methods of alternative medicine, including acupuncture, including the new variant of it offered by our synergeticists, are "soft" and so often less harmful than the "hard" techniques. This is particularly so since the clientele of the alternative methods, unlike the clientele for traditional medicine, are people who despaired of established medicine, so that there is no serious risk due to some mis-diagnosis of an illness which urgently invites some surgery.

I apologize for the lengthy digression into alternative medicine, but I felt that this is some vindication of our synergeticists, and I felt a need to be as conciliatory as I can – not enough, I am sure, but it is the best I can, since I cannot honestly endorse their claim for scientific status for their techniques. If our synergeticists are not discouraged by the ungenerous reaction of some of us, myself included, and it they advance some new ideas, the we all may meet again in some regional AFOS conference in the near future, and with less frustration and more exchange.

Let me close this presentation by reminding you that I found it necessary to dwell on the dar side of our workshop, as becoming critically-minded people, but that the bright side was decidedly the more significant and we all come out of the workshop with a positive sentiment, ready to go to the conference in Warsaw and tell the other AFOS people we will meet there that it was a pity that they missed an exciting experience.